School/Law Enforcement Partnerships:

A Guide to Police Work in Schools

ISBN 0-915920-95-6
2001019889
School/Law Enforcement Partnerships: A Guide to Police Work in Schools

First Printing, April 2002

Library of Congress Cataloging-in-Publication Data

Dorn, Michael Stephen, 1961-
School/law enforcement partnerships : a guide to police work in schools / [by Michael Stephen Dorn].
p. cm.
ISBN 0-915920-95-6
1. Police services for juveniles--United States. 2. School violence--United States--Prevention. 3. Students--Crimes against--United States. 4. Schools--Security measures--United States. I. Title.

HV8079.25 .D67 2002
371.7'82--dc21

2001019889

For FREE listing of related security books write:

Ram Publishing Company

P.O. Box 38649 • Dallas, TX 75238

DEDICATION

This book is dedicated to the brave men and women of the Bibb County, Georgia, Board of Education Campus Police Department who put their lives on the line every day to protect the most valuable resource of our community—our children. We thank you.

DISCLAIMER

CONTENTS

FOREWORD

In the wake of recent serious incidents across America, no issue has become more important than the creation of close working partnerships between schools and law enforcement officials. Many school districts, both public and private, are establishing school safety departments or a division of public safety on their campus.

Michael Dorn led one of the nation's most prestigious school safety and security operations. He is an outstanding leader and innovator in safe schools. The material presented in this book, "School/Law Enforcement Partnerships: A Guide to Police Work in Schools" does an outstanding job outlining the roles, responsibilities and strategies for operating a first class police program on the school campus.

Chief Dorn addresses such issues as weapons screening, incident reporting, safe school planning, and the basics of school search and seizure, among a host of other key security components. Each segment provides practical and usable strategies for the school peace officer. Chief Dorn's experience has made him a national expert in school safety and security.

This book is a "must read" for the school administrator wishing to establish a credible and competent school police program. It is without reservation that this publication is recommended for school officials and law enforcement officials nationwide.

Ronald D. Stephens
Executive Director
National School Safety Center

Over the 10 years of Chief Michael Dorn's leadership of the Bibb County Public Schools Campus Police Department, I had the opportunity to watch its metamorphosis from a struggling, marginally trained and poorly equipped group into a first class police unit which now commands the respect and attention of school districts around the country. As a principal of an inner-city middle school in a decaying and crime-ridden neighborhood, I faced daily problems of gangs, an undercurrent of threats and violence, and the public's legitimate fear of entrusting its children to the care of the public schools.

With a growing number of our youth already in private schools and the morale of public educators sagging, our district's superintendent appointed Michael Dorn as Campus Police Chief. Upon his appointment, he was charged to restore order to our school campuses and to reestablish the communities' confidence in our ability to operate safe and orderly schools. Within a short time, Chief Dorn implemented a rigorous training program, revitalized the campus police force, and worked with the school board, the community at large, and other law enforcement agencies to change dramatically the learning environment in our schools and the physical safety of our students. Under his leadership, our schools became not only a model for other school districts, but a source of pride as well. I have seen the methods described in this book at work to create what is nothing short of profound change. If such significant improvement can occur in Bibb County, it can be achieved anywhere.

Dr. Martha Jones Ed.D
Principal
Miller Core Knowledge Magnet School
A Georgia School of Excellence
Macon, Georgia

ABOUT THE AUTHOR

Michael S. Dorn is responsible for research and development pertaining to school safety for a large southern state. Prior to is work in state government, he was a campus police practitioner for 20 years. He was appointed Chief of Police for the Bibb County, Georgia, Public School System at the age of 27 and served in that capacity for ten years. The department grew to 28 sworn personnel during his tenure. The department was widely used as a model program by organizations including the U.S. Department of Education and the U.S. Justice Department.

Dorn holds B.A. and Master's Degrees from Mercer University. He has completed more than 2,700 hours of police academy training and is a graduate of the 181st Session of the FBI National Academy at Quantico, Virginia. He holds five advanced Georgia peace officer certifications. Georgia State University selected Dorn for an international fellowship program. As part of the Sixth Annual Delegation of Law Enforcement Executives to Israel, he received extensive training in Israel by the Israel National Police in 1998.

The author has made presentations before more than 30,000 education, public safety, and government officials in more than 30 states. He has instructed in such prestigious settings as the FBI Academy and the Oklahoma Governor's Mansion and at more than 100 state, national and international professional conferences. He also developed and presented a series of school weapons violence reduction seminars across the country for the International Association of Chiefs of Police following the school shooting in Jonesboro, Arkansas. Dorn has provided consultation and technical assistance to more than 1,500 school systems, law enforcement agencies, and other government organizations. He has been regularly interviewed by national media representatives in matters relating to school safety.

Dorn has served on a number of national focus groups and participated in U.S. Attorney General Janet Reno's task force to reduce gun violence. He has received numerous professional awards and recognitions during his career. He was selected as the Law Enforcement Officer of the year by the National Rifle Association for his efforts to reduce gun accidents among school aged children and was honored by a resolution passed by the Georgia State Senate in 1999 for his work in reducing school violence.

Dorn is widely published with more than ninety professional articles, manuals, guides and government publications relating to school safety. His articles have appeared in *Police, Police & Security News The Police Chief, School Planning and Management* and *Law and Order* Magazines. He writes monthly columns for *School Planning and Management, Campus Safety Journal* magazine and was a contributing author for a textbook on community policing for Prentice Hall. Dorn has been featured in seven different school safety videos now in use in more than 20 countries and has assisted in the development of several others.

The home search technique developed by the author for the school setting has now been used to successfully thwart several dozen planned school shootings and bombings across the United States.

Visit the author's web-site at www.safehavensinternational.org.

INTRODUCTION

In recent years, I have had the opportunity to reach out to thousands of educators, local, state and federal law enforcement officials, and other elected officials at training seminars and conferences around the country. I have had the privilege to share information with countless other professionals through videos and various publications. It is my hope that these skills and techniques that I have developed have touched the lives of children that I will never meet. Yet, this book will enable me to reach out to an even greater number of people.

I have undertaken this work in an effort to give back something to a field that has been very good to me. In the following chapters, I recommend various programs, training and techniques. Readers should be advised that I do not receive monetary compensation from any of the organizations or individuals mentioned. Although I am currently only 39 years old, I have already enjoyed an extremely pleasant and rewarding career as a campus police officer. More importantly, I want to be able to have the same positive impact on school safety around the country that I have had in my home state of Georgia. If our children are to grow up to be healthy, happy, productive, and gainfully employed, our schools must meet today's challenges. If our children cannot be safe at school, they cannot learn effectively. I want to be a significant part of meeting the challenge of protecting our nation's youth. I urge you to join me on that quest.

The success of those techniques outlined in this book has been proven by the fact that they have reduced student weapons violations by an estimated 90% in a school district of 25,000 students. The 28 sworn officers and six civilian employees of the Bibb County Board of Education Campus Police Department have directly thwarted six planned school shootings (all gang related), one planned school bombing, and one planned double suicide during the past ten years. The information that this book provides is not theory, but is a knowledgeable guide for practitioners written by a practitioner. While no system can come close to providing an absolute level of safety, the methods outlined here have greatly enhanced the level of safety in the Bibb County Public School system.

Yet, readers are urged not to think in terms of trying to copy the overall program that is outlined in the text, but to create a custom-tailored approach using those specific strategies that are appropriate for each local situation. The most effective school safety programs are those that are carefully designed with a great deal of community participation to reflect the needs of local schools.

It is important to understand that the Bibb County Model was developed through intensive interaction with all key elements of the local community. Countless meetings have taken place involving school police, school administrators, teachers, system employees, local public safety officials, court personnel, probation officials, mental health workers, media representatives, parents, clergy, members of the community and, most importantly, our students. A community policing approach is a never-ending process that relies heavily on numerous collaborative efforts. The Bibb County Public School System's safety program receives incredible support from students, parents, system employees, the media, and the community at large.

Readers will note that I have not addressed many key components of a comprehensive school safety strategy such as peer mediation, teen court, conflict resolution, alternative education programs, early mental health intervention, character education and many others. These efforts are a very important part of the Bibb County Model, but I do not discuss them in this book because I feel they have been adequately addressed elsewhere. I have chosen instead to focus on components of school safety that are not as widely known and have been of great interest to participants in my seminars around the nation. At the same time, the reader should make no mistake regarding the value of these types of measures in providing a safe school environment.

Join me now as we take a no-nonsense look at how we can turn back the clock to the day when parents were not preoccupied with worries about their child's safety when they sent them off for another school day. Follow along as we see how thousands of brave men and women who report for duty each school day can make a real difference in our nation's future. Most importantly, don't be afraid to join the fight to effect change for the better in every way that you can—our children and educators deserve it.

Michael S. Dorn

ENDORSEMENTS

This is a thorough and well thought out book on police work in the school setting. Used properly by both school police and educators, this book could bring about an excellent cooperative and proactive approach to school safety. I think this book should be required reading for all school police and educators. I wish this book had been around earlier in my twenty-three plus years in school police work.

Noel E. Zeiser
Washoe County School District Police Department
Reno, Nevada

The *unthinkable* happened at Frontier Junior High, in Moses Lake, Washington, on February 2, 1996. A 14-year-old student brought two handguns and a rifle to his algebra class and killed two students and a teacher. Another student in the class was badly wounded. The Moses Lake School District is a very "normal" school district of 6,000 students in the center of the states' agricultural heartland. There has always been excellent community support for operations levies and school construction bonds. The academic and extracurricular programs are strong. However, the unthinkable occurred here.

Schools in the Moses Lake School District were perceived by parents and students to be safe prior to February 2, 1996. The schools are perceived to be safe now. The district had several safety programs in place prior to 1996, and has added more in the past few years. One of the school safety and security experts we have learned from has been Michael Dorn. He has taught us a great deal about how to improve safety and security in our district. This book describes many of the things we learned from him. As he states in his book, "what works in one school setting may not work, be legal, be adequate, or may be "overkill" in another school setting". We have not implemented everything he offers as possible components, but we learned many practical lessons from him on how to improve safety and security. If you are a school administrator, this book will help you improve safety and security in your school too.

Steve Chesnut, Ed.D.
Superintendent
Moses Lake School District

This book is based on first hand information obtained in a real school system. The success story of the Bibb County Public School System in the reduction of violence is a result of the strong desire on the part of the campus police, parents, students, faculty, staff and the school board to create a safe environment for young people to learn. The "how to" techniques to reduce violence that are contained in these pages are practical ones. The police, and school administrators can take what they believe will work and customize an approach to fit the individual communities and political and social situations they face.

Paul Bolton
International Association of Chiefs of Police

I was requested to review the manuscript of *School/Law Enforcement Partnerships: A Guide to Police Work in Schools*, written by Michael Stephen Dorn. The manuscript is well written and I would consider it to be a comprehensive guide to law enforcement officers and security personnel in our present day issue of violence in the school system.

The book expertly deals with issues of police and security services in the schools, including the screening, selection, and training of them. Effective school weapons screening (firearms, knives, etc.) and campus patrol methods. Basics of school search and seizure, and investigative techniques are dealt with as well as dealing with youth gangs and violence in the school setting. United States Supreme Court rulings are also included in the manuscript. The manuscript is logically planned out and contains an enormous amount of practical examples, strategies, and solutions.

It is clear that Mr. Dorn is an expert in this field and through his experience and teaching, he will affect the lives of many children. There is a quote that I feel is appropriate and that best fits his manuscript, "Quality is never an accident, it is always the result of high intention, sincere effort and intelligent direction. It represents the wise choice of many alternatives"

We have the opportunity to curtail and prevent the spreading of violence, an issue of growing concern here in Israel. I recommend that his manuscript be translated to Hebrew and that Mr. Dorn be invited to Israel to lecture and teach.

Howard Silverwater, Superintendent
Forensic Firearms Identification Lab
Division of Identification & Forensic Science
National Police Headquarters
Jerusalem, Israel

Chapter 1

POLICE IN THE SCHOOLS

Across our great nation, in small towns and in major urban centers, increasing numbers of police officers are reporting for duty in America's public and private schools at the secondary and elementary level. Thousands of police officers now work full time as school resource officers or school district police officers. In New York City alone, 3,500 sworn officers are assigned to the public school system. In this book, the terms "School Resource Officer" and "Campus Police Officer" will be used interchangeably, and "School Systems" will refer to public and private elementary, secondary schools and school systems. Across the nation, there is often a great deal of overlap even though differences exist between the two in specific regions.

The trend to place police officers in the school setting is following a similar course that took place at institutions of higher learning in the 1970's and 1980's. While several elementary/secondary school police units are 25 years old and the Flint Michigan Police Department began assigning sworn officers in the schools in 1957, most school police officer positions in this country have been created in the past 10 years. For example, in the state of Georgia, only 12 school systems had sworn personnel assigned in 1989. A recent survey revealed that 141 of 159 public school systems, and a number of private schools, have a school resource officer program. Police assigned to the schools have become common across this state.

Depending on which study is considered, it is estimated that between 100,000 and 250,000 students each day carry a gun to school. With an estimated 18 million student gun violations each year and less than 4,000 guns being reported as recovered to the U.S. Department of Education as required by law, it is estimated that only one gun is recovered for every 4,500 student gun violations. The National School Safety Center maintains a list of violent school deaths in the U.S. since 1992 based on media reports. The list now includes more than 270 violent school deaths at public and private schools in our nation. It is clear that there is a need for law enforcement assistance in our nation's schools.

Police officers are increasingly being utilized in schools as the idea gains

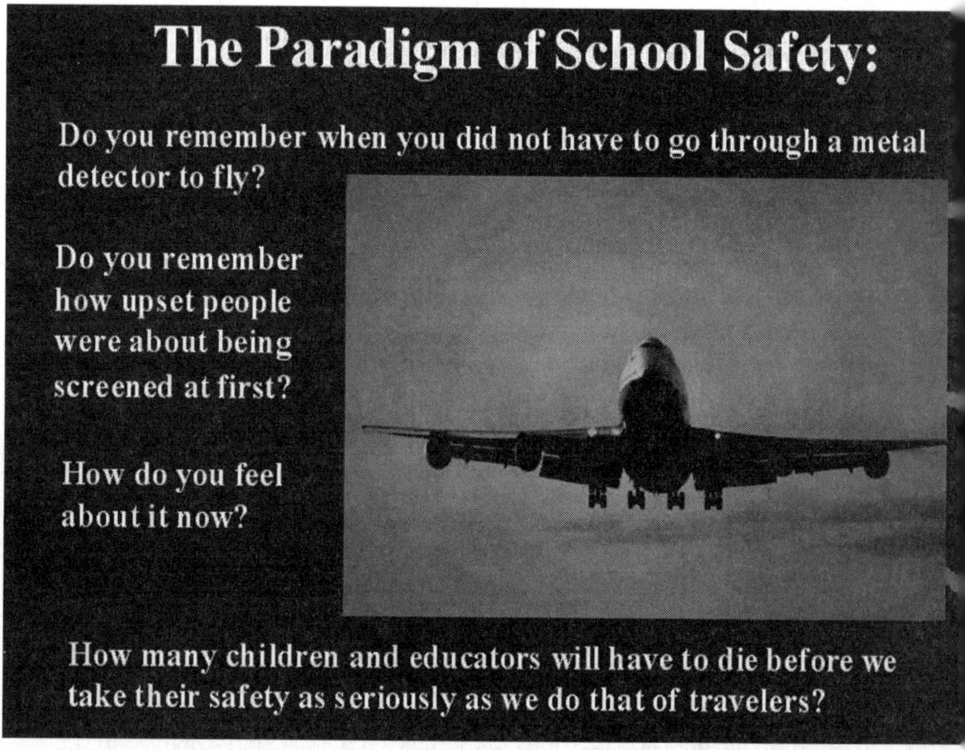

The author uses this analogy to help participants in his seminars reexamine the issue of school safety. How many acts of violence will have to take place in our schools before every school develops an effective prevention strategy? Passengers choosing to fly are provided with a level of protection while our youth who are compelled by law to attend school are often not afforded this level of safety. While different measures are needed for schools than for airports, the level of commitment should be present in every school.

acceptance and popularity with school administrators, school boards, students and the general public. The increasingly frequent and often substantial costs of successful litigation against school systems that experience a weapons assault and do not have officers assigned on campus is convincing many systems to add police officers to the school staff. More and more, police officers are becoming an accepted—even expected—part of the school environment. Police officers are now considered to be a critical component in many schools. Some teachers and school administrators have told the author that they would retire early or change fields if they had to work without officers assigned to their schools.

Law enforcement officers are becoming a common sight on and near campuses. Here Bibb County School District Police Officers arrest a drunk driver who had a juvenile female student in his car and was attempting to return her to the school where she was truant.

Police officers, who are carefully screened, extensively trained, and properly utilized, can make a tremendous difference in a school setting. In fact, school officers often find that they can make more of a difference in a school setting than in traditional police assignments. An active school resource officer program can improve grades and reduce truancy among students. Such a program can also help to reduce stress, absenteeism and turnover among school staff.

School employees and students typically respond very well to a quality school police program. Police officers often find assignment to schools to be rewarding even though the assignment can be more difficult and complex than traditional patrol work. There is an adage among veteran street police officers who have worked in a school setting that if officers can effectively police in the campus environment, they can police anywhere.

All actions of a school police officer should be directed at maintaining a calm and peaceful environment so teachers can teach and students can learn. School police officers, therefore, must be flexible, patient and, most importantly, keep a proper focus on the actual level of safety at the school. The campus police role is to minimize disruptions to the learning process on and near the campus.

There can be no question that active school police personnel can significantly reduce crime on campus. Of primary concern are crimes of interpersonal violence, gang-related crimes, weapons, alcohol and drug-related offenses. Unlawful disruptive behavior is a less dangerous but more frequent problem in many school settings that can be addressed by campus police. There are a host of other violations that are also handled by school police personnel.

School safety zone enforcement can be a very important task for officers who want to increase the feeling of safety within their assigned schools. Crimes that occur near campus are just as important as crimes on campus. A case of a student being shot and killed across the street from a high school can create fear and unrest that can effectively hinder the learning process. Cleaning up the neighborhood around the school enhances the actual and perceived level of safety in the school. Heavy traffic enforcement and enforcement of loitering statutes are effective tools in cleaning up these areas.

Through educational programs and positive interaction with students, school police officers can improve police/community relations and have a more positive impact on community crime problems. Indeed, well designed programs will reduce the number of young people entering the criminal justice system as criminals or victims. Officers assigned to schools should see one of their major roles as helping students learn to succeed in life as adults. Officers who have this mindset can have a powerful impact on young people at their assigned schools.

It seems clear that police officers have a major influence on our schools. There are several critical issues to examine when assigning officers to schools.

1. Who will bear the expense of placing police officers in the school setting?
2. Will the officers be armed?
3. Under what circumstances will officers make arrests?
4. What areas have demonstrated the most critical need for topics for educational programs?
5. What means will be utilized to prevent violations concerning weapons, drugs, and, alcohol?

Unless these issues are resolved, problems can arise that could easily have been avoided.

Several approaches are commonly used to cover the expenses associated with police coverage in schools. One method is to secure funding through a grant. Partial grand funding has enabled a number of successful programs to begin. Grants for police in schools frequently cover 25% to 75% of the cost of the project. The obvious drawback to this method is the fact that grant funding normally only lasts from one to four years.

The Bibb County Board of Education Campus Police Department received nearly one million dollars in federal grant funding in a two-year period. Deputy Chief Russell Bentley was able to obtain these grants for the district by carefully documenting incidents and problems in the district and by properly articulating how the grant funding would be effectively utilized to improve the level of safety in the schools. Schools often have difficulty obtaining grants because they fail to document and report incidents properly and, thus, cannot demonstrate need for the funding. Another problem is that many districts are not aggressive in researching and applying for grants and neglect excellent funding opportunities.

Another method to fund a school police program is to defray costs by charging a parking fee to those students who drive to school. Because one of the primary duties of the officer is often to patrol the parking area, students who pay for the service receive a significant portion of the patrol services rendered. A number of schools use this procedure to maintain police coverage. One advantage to this method is that it provides a continual source of revenue to fund the program.

The majority of school police programs are funded by the school system, a local police agency or a combination of the two. Commonly, local city police, sheriff's deputies or county police personnel are assigned as school resource officers with the cost split between the police agency and the school system. Frequently, the school system bears the bulk of the expense. In larger school systems, it is often found that the school system can operate its own district police force in a more efficient and economical manner than utilizing local officers on contract.

The issue of arming school police officers can be a touchy one. After careful study, years of experience and discussion with hundreds of school police personnel, the author's firm opinion is that police officers who work in a school setting should always be armed. It is my opinion that officers should be equipped with a standard issue duty firearm (with standardized ammunition), a high security retention holster, pepper spray or mace and an impact or restraining device. There have been two fatal attacks on school district police officers in Georgia, one in Florida, one in California, and one city police officer fatally shot in an Illinois school. There have no doubt been other cases beyond the author's knowledge. Unarmed officers are helpless to prevent death and wounding against an armed individual. Several incidents committed in an officer's presence starkly illustrate this point.

Police work is dangerous in any setting. While statistically rare, shootings, stabbing, murders and hostage situations have occurred in our nation's private and public schools. These situations happen in the most unexpected places including the secondary and even elementary levels. A police officer without the availability of various force options is of extremely limited value to the school community. Not only can he or she not effectively deter a planned weapons assault but can often become yet another victim of assault.

Several other problems are encountered when unarmed police officers are assigned to the school setting. First, many unarmed school police officers admit that they "hold back" a great deal in the performance of their duties because of their extreme vulnerability and lack of adequate protection. An unarmed officer will sometimes ignore obvious violations because of the clear risks that intervention can pose. A second problem is that a number of "unarmed" officers routinely carry a small handgun in a back pocket or ankle holster. In one such case, a Georgia officer had the unpleasant experience of his unauthorized .25 caliber semi-automatic pistol discharging in his back pocket while he was on duty in the school. This situation is problematic because these officers are not necessarily trained and range-qualified with these weapons and, as in the case above, may choose a weapon and method of carry that is unsafe for police use.

A third danger is the increase in the use of force that is experienced when unarmed officers are utilized. Offenders resist arrest and assault unarmed officers at a much higher rate than officers who clearly have the means to defend themselves. Officers who only have the option of open hand control and restraint will use force more often, face more injuries and will injure the suspect(s) more often than those with proper equipment options. Finally, it can be very difficult to attract, hire and retain quality law enforcement personnel to work in an unarmed police assignment.

An equally touchy issue is determining when the officers will make an arrest. School administrators are held accountable for what happens in their school. Principals are typically in charge of anything that takes place in the school facility. With this in mind, we can better understand why school administrators often do not like to relinquish authority for arrest decisions to police personnel. Likewise, police officers are not accustomed to having a civilian make arrest decisions for their criminal cases. Both administrators and police officers feel that they will be held accountable for decisions that are made by the other party.

Some school systems allow the building administrator to make most arrest decisions. A major drawback to this concept is the risk of civil and criminal liability. If a school administrator makes a legal error and gives improper instruction to a school resource officer who has been ordered to follow the administrator's instructions, the administrator, the police officer, the school system and the police department may all be held liable in civil court. Similarly, if a code violation, such as false imprisonment under color of law takes place, both the administrator and the officer may find themselves under indictment and facing criminal charges. Unless the school system is willing to invest weeks of staff time for proper legal training of each administrator, allowing them to control arrest decisions is usually best avoided.

Other school systems leave the arrest decision-making authority to the school police officer. Although this arrangement is similar to what officers face in traditional assignments, it may not always be appropriate in a school setting. The original concern of school administrator is evident-they may be held accountable for the actions of the school police officer without having the authority to influence performance. Though this method may improve the liability risk for all involved, it does not hold the administrator and school police officer properly accountable for school safety, even though each has been given the proper authority to act.

A third possible scenario is for the school board to establish written guidelines for the arrest of students. If the school board members are adequately briefed on problems faced at the building level, workable guidelines can be established. One example of effective guidelines are those implemented by the Bibb County Board of Education in Macon, Georgia. These guidelines give direction to students (all students receive a written copy), school administrators and district police officers. This type of policy helps to ensure consistency within the school system and, most importantly, reduces violations because everyone is aware of the consequences.

Pages 11-12 of the Bibb County
Public Schools Student Discipline
Code of Conduct 1997-1998 states:

"CRIMINAL PROSECUTIONS"

In addition to the normal school discipline procedures to be followed under the Student Code of Conduct, misconduct which is also a felony under the laws of Georgia or the United States, or is a delinquent act which would be a felony if committed by an adult shall be reported by the principal to the Campus Police Department for prosecution.

Less serious criminal conduct as listed below shall also be reported for investigation by Campus Police:

Fighting
Battery (including Sexual Battery)

Disrupting a Public School
Criminal Trespass
Possession of Drugs
Loitering in a School Safety Zone
Possession of Alcohol and Possession by Consumption
Traffic Violations
Misdemeanor Theft
Damage to School Property (writing gang graffiti, etc)
Pulling a Fire Alarm
Bomb Threats
Inciting to Riot
Unlawful Assembly
Contributing to the Delinquency of a Minor
Gambling

Except in cases where school safety is a concern, the foregoing prosecutions shall normally be handled by means of a citation to appear in Municipal Court, the State Court of Bibb County, or the Juvenile Court of Bibb County.

The above guidelines have proven effective in reducing incidents, creating a calm environment, providing consistency from school to school (the district has eleven secondary schools) and establishing a fair and equal set of guidelines for all students and staff. Some might feel that these mandatory guidelines are too severe. It is worth considering, however, that it is more unfair to send mixed messages to the student body by not having clear and consistent guidelines. Most students will conform to the rules of the school system if they are informed clearly about the rules and they can observe that violators are consistently punished in a firm manner.

A key component of this approach is the use of citations to appear in court to deal with minor criminal offenses such as fighting. The use of citations impresses upon students the seriousness of the behavior while avoiding the use of a custodial arrest. This approach can allow fines and community service hours to teach students, parents and the community that the school district is serious about maintaining safe and orderly schools. In one Texas school district of more than 30,000 students, fights at school became so infrequent that the superintendent was notified whenever a fight occurred.

When we look at the more common "triggering behaviors" that typically precede school weapons assaults, it becomes more apparent that the traditional methods that have been attempted over the past decades are no longer adequate for our changing society. When parents and students understand that these behaviors will result in consistent consequences and that the methods designed to control such unacceptable behavior are in place to prevent more serious incidents, the level of violence and disruption in the school will decrease significantly.

Educational programs taught by school police personnel should be carefully matched with the most significant problems faced in the school system and the community at large. Officers teach programs on many different topic areas such as drug abuse, dangers of gang activity, conflict resolution, weapons safety, traffic safety and law-related education to name just a few. The key is to identify the critical community needs and develop a quality program that is politically neutral in content so that it may be taught without distraction and with broad public support. Great care must be taken to avoid programs that have a clear political message. Careful consideration must also be taken when designing programs that contain any discussion of a sexual nature such as a rape prevention program. Many school system policies and state guidelines clearly spell out parameters for discussing information of a sexual nature in the school setting.

Careful planning, open communication and a good dose of common sense can eliminate most pitfalls in the selection of topics for educational programs taught by school police officers. Benefits to the school, students, police and the community at large greatly outweigh the obstacles of implementing a quality program.

A needs assessment of topics to be taught by school resource officers may involve a review of available statistical data, meeting with groups of students, parents and staff members or the conducting of surveys. Often, statistical data will reveal critical needs for classroom instruction. An example would be statistical data that reveals that significant numbers of students are being arrested for obstruction of police officers. A quality law-related education program can reduce these arrests as students understand that interfering with a police officer is an unlawful act. Again, the vast majority of students will not "cross the line" if they understand that to do so is a crime and that they can be arrested for such an act.

Group meetings can be very insightful. Discussion often reveals that topics that may not have occurred to the police officer are found to be of great interest to the students or faculty. Surveys can be very productive, but care must be taken to ensure that they are valid, and that school system policies, state guidelines and federal statutes for student surveys are not violated.

Developing an effective strategy to deter violators from bringing contraband onto campuses is of considerable importance. The role of the school police officer in contraband reduction is a vital issue. Will the police officer(s) be involved with weapons screening such as metal detection, locker inspections, and plain view inspections of student vehicles? The most effective screening programs involve uniformed police personnel. Guidelines should be established to direct officers on their responsibilities within the weapons and contraband screening program.

The police role in school contraband screening is crucial. Most parents and school personnel would be astonished at the amount and type of contraband that is brought into every school daily by students and visitors. Typically, the greatest shock would be in smaller school systems and private schools. In one study, it was determined that more than 300 incidents of firearms being discharged in American public and private schools took place in one year. Although assaults with other types of weapons are far more common, the number of weapon assaults in our schools is cause for serious consideration. Other types of contraband also cause a great disturbance in the education process in our schools. Experiences of educators from around the nation have demonstrated that it is reckless for school officials themselves to search students for weapons. In case after case, weapons are missed or educators are injured, killed or taken hostage when they attempt to perform a task for which they are not properly trained or equipped.

Effective countermeasures for dealing with contraband on campus in this day and age must positively, without a doubt, involve some type of screening program. While contraband screening may be distasteful, it is a fact of modern life. Other options are even more undesirable. Airline terminals, courthouses, public buildings, and even police departments now routinely use metal detectors to screen visitors. In each of these settings, current trends have mandated a response designed to protect the public. In a school setting, students have legal rights relating to their safety that do not exist elsewhere in our society. The courts have had little patience with school systems that do not take these rights seriously.

As police personnel become a part of the educational setting, we can expect an increasing number of specific challenges to be addressed by these officers and the school administrators who share the responsibility of keeping our children safe. Each school system must make decisions regarding the actual extent of the police role. The general considerations that have been discussed should be helpful in establishing viable answers to meet local needs.

The remainder of this book will discuss a variety of options for the successful utilization of law enforcement officers in the school setting. It is not suggested that specific techniques are right for every school. Rather, school districts should focus on creating a custom-tailored concept that is appropriate for their community. Readers should view specific methods that will be discussed as items on a menu for consideration. Each school and each community have similarities but the differences between them dictate a locally designed approach that will address the problems faced by all schools while focusing on those that are unique to each community.

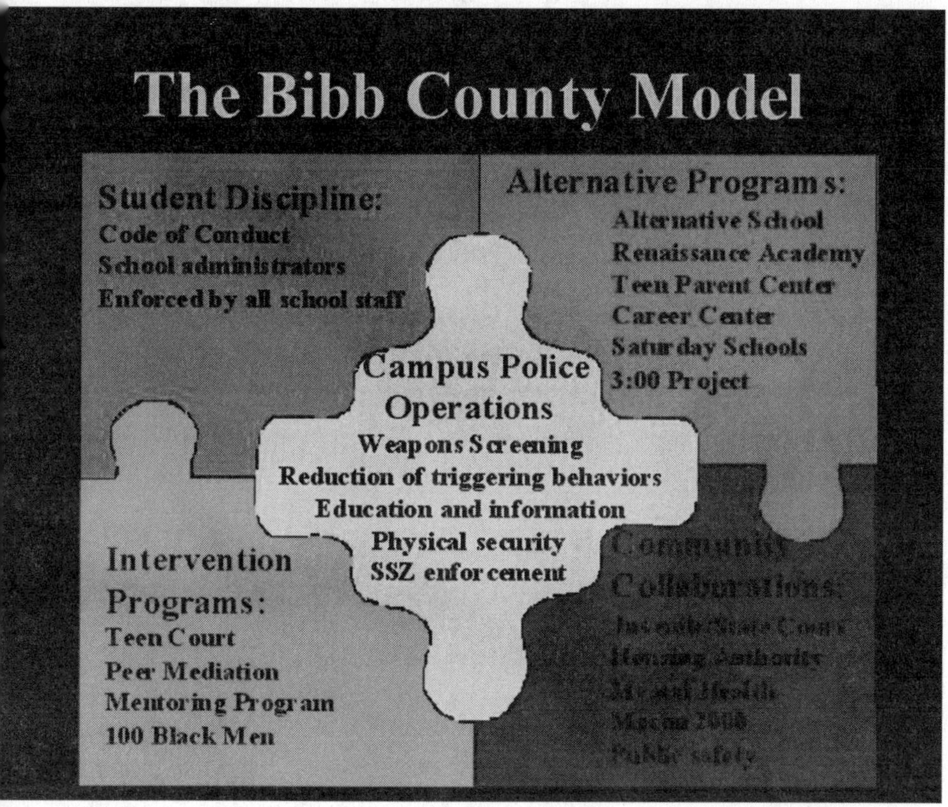

It is important to understand how law enforcement efforts are only one component of a comprehensive school safety strategy. While this graphic lists a few specific examples, the Bibb County model actually relies upon more than 100 individual components which are utilized cooperatively to address the complex issues of school safety. The individual techniques are not as important as the concept of a community-based approach that has been tailored to fit the needs of the local community. Each community and each school has unique challenges and different resources. In the Bibb County model, the department's gun detection dog is no more or less effective than the peer mediation program. By combining a number of specific approaches cooperatively, significant and lasting improvement can occur. While this text focuses on law enforcement approaches, the reader should be reminded that other methods are also of great importance.

Chapter 2

SCREENING, SELECTION AND TRAINING OF SCHOOL POLICE PERSONNEL

A key component of any school resource officer program is the quality of its human resources–its people. Screening, selection and training issues have become increasingly important not only as citizens demand more efficiency from public safety organizations and school systems, but as civil liability issues demand attention. An efficient screening, selection and training system is vital as the foundation for a successful, efficient program.

Screening, selection and training of personnel is of vital importance. The bottom line is "quality in, quality out". To have a viable school police unit, you must have quality employees. There is an old saying in management "if you hire bums and train them, all you have is trained bums".

First we should examine qualifications for officers detailed to work school assignments from city or county police units. These officers have normally completed a thorough screening process to become an officer in a law enforcement agency. It is important, however, to carefully review the officer's personnel file. Has the officer adapted well to the police role? Is the officer productive? Are the number and types of citizen complaints normal for the officer's arrest rate and assignment? Are performance evaluations indicative of a high caliber officer? If these questions are answered in a favorable manner, supervisors should be interviewed to for their opinion on how the officer will adapt to working in a school setting. Finally, the officer should be interviewed to ensure that he or she really wants the assignment and understands what it entails, as well as to assess the officer's outlook on the overall school resource officer program. The officer's philosophy on police in the schools must be in line with the outlook of the police department and the school system leadership.

In many cases, municipal or county officers are reassigned from other duties to work as SRO's. In most of these cases the officers were properly screened when they were first considered for appointment as an officer. But fitness for duty and background screening are only part of what must be considered. Assuming that officers have been carefully screened and selected as police officers, we now must look at how well they match up to an SRO assignment. Some superb street officers will fail in an SRO assignment because they are not suited to work in this specialized setting.

Only individuals who can successfully complete a thorough screening program should be granted the trust that is needed for an assignment involving schools. Bibb County Campus Police Officers patrol a student parking lot.

Two major problems are off-duty employment situations and the SRO program being used as a "goof-up" squad. In the first scenario, the SRO assignment is an extra off-duty job for police officers. This normally results in a poor quality program. With officers working 40 (or more) hours in their primary role and then an additional 40 (or more) hours as an SRO, fatigue can be a serious concern. Officers typically desire this type of off-duty assignment because it greatly enhances the officer's income.

The problem here is that while the extra income can be a positive motivator, a fatigued officer is being being relied upon to protect students. Another consideration is that in many situations of this type, the officer is only allowed to work three days each week in the school setting (this is still a 64 hour basic work week) which also means that the school loses the consistency of having the same officer in place each day.

A fatigued officer will often be at a lower level of alertness when performing critical functions such as:

- working at a metal detection screening station
- conducting random weapons screening
- teaching a class
- making a traffic stop
- making a use of force decision
- driving a car

Some of these are routine, daily duties. Use of force decisions may be rare in some settings, but this is a critical area with little room for error. All of these duties must be performed correctly. Top performance in any area requires an alert officer.

Another problem is the use of an SRO program as a unit for placement of "trouble" officers. In some cases, officers who are lazy and incompetent or exhibit some other significant deficiency are placed in an SRO assignment to "get them off the street." This is potentially a tragic situation waiting to happen. Units set up along these lines will fail to provide adequate services.

An SRO unit should be (and often is) a prime detail with more qualified volunteers than available assignments. Anything that the law enforcement agency or school system can do to make the assignment desirable will help ensure quality personnel and, thus, an effective program. In many Georgia communities, school resource officers receive significantly higher salaries for this reason.

SCREENING

At the beginning of the application process, all candidates should be made clearly aware that anyone who is found to be untruthful in any way during the screening process will be immediately disqualified. A police officer has the ability to take away the freedom of other citizens based on arrest and courtroom testimony. Persons who will be untruthful when seeking employment cannot be trusted with this awesome responsibility.

A quality personnel selection process will significantly enhance the likelihood of a quality school safety system. Adequate screening of candidates is a cornerstone of establishing and maintaining a quality school enforcement unit. A comprehensive process should be utilized. Some steps that are often appropriate include:

1. Complete criminal history check - Any arrests that are disclosed during this process should be carefully investigated. An applicant's version of any suspicious incident should not be relied upon. If available, the police incident report should be reviewed and the arresting officer should be interviewed. It is important to remember that people are often charged with minor offenses when they have actually committed a much more serious crime i.e. simple battery when the suspect has actually committed a rape. Often, officers will charge an offender with a solid misdemeanor charge rather than a weak felony charge. The same is true with conviction information since it is very common for serious felony charges to be reduced to a misdemeanor charge in exchange for a guilty plea. At the same time, it is important to realize that an arrest of a minor nature should not always automatically eliminate a candidate. There are cases where innocent people are arrested and it is just as important to give such individuals fair consideration as it is to screen out unfit applicants. One caution on this point is appropriate: applicants who have more than one prior arrest will rarely be good police material. This is particularly true when the charges are not related to the same complainant or situation.

2. Complete driver history check - In this area again one ticket may or may not indicate a problem. Candidates with a series of tickets may be reckless and irresponsible. Particular attention should be paid to serious citations such as driving while under the influence (often indicates a substance abuse problem), leaving the scene of an accident (usually indicates a lack of integrity and may be related to a person attempting to avoid arrest for driving under the influence).

3. Require that all applicants complete a thorough background check booklet - *(See example p. 141-174)* The booklet should include questions about work history, disciplinary actions, terminations, arrests, traffic tickets, past drug use, theft, other criminal acts, mental problems, financial problems and any other areas of concern that can legally be explored relating to honesty, integrity, reliability and stability. The applicant should also list information regarding neighbors on all sides, including phone numbers and addresses so they can be contacted. Answers in the booklet should be reviewed and compared to other information such as polygraph and integrity test admissions.

4. Polygraph test - This is a very critical part of any effective screening process. The quality of the examiner is of paramount importance. While many people question the accuracy of the polygraph, the startling number of applicants that are washed out of the screening process based on admissions to serious criminal behavior during polygraph testing alone make the use of the polygraph worthwhile. (Note: there are other types of instruments currently in use that can serve as an alternative to the polygraph).

5. Integrity/honesty test - As long as a reputable company and valid test are used, these tests can also be a practical way to screen out people who do not have the strong character that is needed to be a police officer.

6. Credit history check - Individuals who cannot responsibly manage their personal finances typically experience problems trying to manage the position of police officer and frequently fail. Be sure to check the law in your state to see if such information can be considered when screening applicants.

7. Drug screen - This can often weed out people who clearly have no business in the field. Special attention needs to be paid when positive test results indicate the use of unauthorized prescription drugs. More police applicants are commonly weeded out due to abuse of prescription drugs and steroids than for "street drugs."

8. Psychological testing - An appropriate written test instrument accompanied by an interview with a psychologist can be another important screening phase.

9. Written reports from at least six work-related references - A reference packet that has specific questions that are relevant to the job of police officer should be obtained and reviewed, with each reference contacted for confirmation and additional comments.

10. Interviews - Carefully conducted interviews that are designed to evaluate the candidate's verbal skills, work skills and aptitude for quality service should be mandatory for all applicants. Interviews should be structured to concentrate on relevant information only. This will help to ensure that legal problems do not arise and, more importantly, to ensure that the best candidates for police positions are selected. A series of questions that require candidates to provide examples of past work performance are very useful.

Questions that require the candidates to give examples of instances that the candidate feels in their own assessment illustrate situations that the candidate handled extremely well and situations that the candidate feels that they should have handled better tend to provide valuable insight. These types of questions are more reliable indicators of how a candidate will perform than questions along the lines of what would you do if...? As with background investigation information, the best indicator of future performance is past performance.

Careful consideration to negative comments made by candidates about their current or prior employers should be given. Very often, candidates who view these employers in a negative way tend to develop a negative view about future employers as well and may become "problem employees". It is also important to try to determine if candidates have a normal and healthy view of themselves. Applicants who do not have well adjusted view of themselves and others will often experience difficulty in a school police role. It is of paramount importance that great attention be focused on assessing why the candidate wants to serve in a school police assignment. It is very important not only to screen out those candidates who will not be able to perform well, but to screen out those candidates who are capable but who will not be content in the unique position that they have applied for.

A candidate who accepts a position only to leave six months later is expensive to the organization and may not perform properly as well.

SELECTION

A panel of officers (and possibly school administrators, teachers, parents and students) who conduct a structured group interview for each candidate can be very helpful in effectively screening applicants, and in encouraging officers to support new officers in the department. Line officers can assess candidates from a different perspective than management, which can be important in securing new officers who can work well as a team with existing staff.

If line officers have a significant role in the screening and selection of new officers, the rank and file will be more supportive in "breaking in" new officers than with officers chosen by "the brass" alone. Assessment panels have also been well received by the courts as a screening process in a field where judgment, problem solving and communications skills are critical. It is, of course, absolutely necessary that members of the assessment panel receive proper instruction in order that legal and effective assessments can be conducted. Many states have structured assessment panel training programs available and in some areas, panels are convened to screen applicants for other departments.

Not utilizing a thorough screening process can cause long-term problems with unfit personnel that can be difficult to resolve. Attention to past behaviors of candidates for school law enforcement positions should be a prime consideration. Most veterans who are experienced with police officer selection agree that past behavior is the best indicator of future performance.

TRAINING

Once the right people are selected, they must receive a solid training background to maximize their effectiveness. In the specialized school environment, the training process is not a one-time proposition. A solid training program will result in continually improving the capabilities of each officer.

There are many areas of specialized training that will enhance a police officer's effectiveness to protect students and employees in a school setting. A few topics that are of particular value to school police work include:

- School search and seizure law
- Juvenile law
- Visual weapon screening
- Principals of proper metal detection
- Recognizing disguised weapons
- Drug identification
- Dealing with special education children
- Advanced use of force training
- Advanced judgmental firearms training
- Recognizing gang indicators
- Understanding cults and violent extremist groups
- Crime prevention
- Lesson plan development and effective instructional methods
- Crisis intervention/critical incident management
- Community policing strategies

Training needs of the school police officer are obviously very diverse. The requirements of a school law enforcement assignment demand a flexible officer who is prepared to deal effectively with a wide array of situations.

Some might question the need for the types of training listed above. To illustrate the necessity of diverse and extensive training, consider the following scenarios:

- An officer is assigned to a school with no gang activity. Several students who are gang members from a larger city transfer to the school when their families move into the community. If the officer is not familiar with recognizing and dealing with gang activity, a full-blown gang (and weapons) problem will probably develop within a relatively short time.
- An officer assigned to any school may suddenly be faced with a critical incident such as a fire, a lightning strike, a tornado or a violent weapons assault. While such incidents are statistically rare, they occur with enough frequency to be of real concern. Emergency personnel must have a solid foundation of knowledge regarding critical incident management before they are faced with it. Trying to learn how to handle a situation as it is happening can be expensive in terms of loss of life and needless suffering.
- A teacher asks an officer to search a classroom filled with students to locate her wallet that was clearly taken by one of the students when she stepped into the hallway. An officer who makes the mistake of searching the students in this scenario is exposing the school system to clear civil liability. Few, if any, courts would uphold this as a legal search upon challenge by a competent attorney.

Screening, selection and training of school-based police personnel comprise a starting point for any effective unit. If even one of these key areas is neglected, a significant reduction in the program's overall quality is certain to result. Proper emphasis on these three areas will be rewarded with worthwhile gains in performance and reliability of personnel.

In the Appendix to this book, the background investigation booklet that was adapted by the Bibb County Board of Education Campus Police Department from a document provided to the department by the University of Georgia Police Department can be found. This booklet dramatically improved the background investigation process utilized in Bibb County.

Be sure to review current laws in your state regarding privacy issues before making changes to your background investigation process. Because of the sensitive nature of the school law enforcement role, it is the author's firm belief and experience that a more thorough background investigation is required for officers who will serve in this capacity than would be needed for normal law enforcement positions.

The responses in the booklet should be compared to the results of other investigative steps such as interviews, criminal history, driver history, credit history, interviews with neighbors, polygraph, and integrity test results. Any apparent falsifications by the applicant should cause the hiring agency to seriously consider disqualification of the candidate.

The booklet also serves to indicate how thorough and how efficient the applicant will be in completing required paperwork if hired by the agency. An applicant who is careless in filling out the booklet to obtain a position is likely to be even more careless in doing the work that the position requires if they are selected.

Chapter 3

INFORMATION AND EDUCATIONAL PROGRAMS

Informational and educational efforts can have a positive effect on the level of safety in the school setting. A quality safety program will ensure that students have an opportunity to gain a clear understanding of school policies and laws of particular application in the school setting. The majority of students will comply with policies and laws if they understand what is expected of them. It is important to understand that information and educational programs are actually two separate but related concepts, both designed to help students understand and determine appropriate behaviors.

Information programs are designed to ensure that students and visitors receive adequate notice regarding school policies and criminal laws applicable to a safe and peaceful school environment. Such programs should utilize a variety of means to put forth the desired information. A good dose of creativity can be very beneficial to an information program.

Some concepts that have been useful to disseminate information to students and visitors include:

- **Code of Conduct** - A well designed, clear, concise and standardized student code of conduct should plainly spell out unacceptable behavior and consequences of violating school policies and criminal statutes.
- **Signs** - Signs concerning preventive search measures and prohibited items such as weapons, narcotics and alcohol should be designed and posted in prominent locations. These signs are usually placed outside the building near parking lot and building entrances.
- **Posters** - Posters utilized like metal and plastic signs can contain more information and are normally used indoors.
- **Behavior Contracts** - Student behavior contracts constitute agreements between the student, parent and the school administrator. These documents ensure that students and parents understand particular policies and consequences of violations. The behavior contracts require that the student and parent agree to comply with and support the policy, and the administrator agrees to enforce sanctions whenever violators are identified (see Appendix A.)

- **Videos** - Concise video presentations can be extremely effective in insuring that students understand what is expected of them. Video presentations often reach those students who may not absorb the message effectively through other means. Video can be particularly effective in schools with classroom television hookup. It is important that video presentations not only be short, but they must also be designed to hold student interest. Video presentations relating to weapons, drugs and fight reduction policies can be effective in gaining voluntary compliance, parental support and more effective prosecutions.
- **Intercom Announcements** - Occasional, brief intercom announcements to remind students of important policies can also be very effective.
- **News Media** - Working with the media to get the word out about weapons and violence reduction efforts can be useful. Of course, great care must be taken since news people frequently put a negative "spin" on such stories and may sensationalize them. Even with the negative factors often associated with news media, involving them to inform students, parents and visitors regarding consequences of violating school policies can be worthwhile.

A host of other informational programs can be developed and are limited only by the imagination. A related issue is the use of various age-appropriate educational programs. In these programs police officers enter the classroom in the role of an instructor. In some school systems students are allowed to take classes for credit that are taught by police officers. Utilizing sworn personnel to teach in the classroom can greatly enhance relations between students and officers.

Officer Kenneth Bronson and "Eddie Eagle" teach students how to avoid accidents involving firearms. Officers from the Bibb County Board of Education Campus Police Department have taught this program to more children than any law enforcement agency in the United States. The department has presented this lifesaving program to more than 50,000 children and serves as a national model agency for this program.

The first step in developing any educational program is to assess the needs of the student body and the school system. This can be done through surveys of students, analysis of criminal incident statistical data, examining statistical information relating to serious injury and death of school aged children, and through informal meetings with students, teachers and administrators.

Once the needs assessment has been completed, programs can be developed or selected to meet those that have been identified. Care must be taken when reviewing available programs to select those that are politically and religiously neutral, culturally sensitive, appropriate for school children and otherwise suitable for a school environment. An apparently useful program can fail because it does not meet the above criteria.

Some examples of programs that have demonstrated success in schools across the country include:

- **Law-related education** - These programs are typically custom designed to meet the needs of students in grades 6-12. Usually one grade level is targeted each year. Officers explain what is required to become a police officer, and what services are provided by local, state and federal law enforcement agencies. Students are taught their rights as citizens to utilize police and court services. The rights of victims can also be addressed in this type of program. Students then receive a brief outline of the juvenile and adult criminal justice systems. Officers explain to students what impact an arrest and conviction can have on their future employment opportunities and the loss of their citizenship rights, such as the right to vote. Students then learn about specific traffic and criminal laws that affect young people. Law- related education can give students valuable information that will help them to make wise choices in life to keep them out of the criminal justice system.
- **Eddie Eagle® Gun Safety Program** - This well designed program has been taught to more than 10 million American children. The program teaches children that guns are not toys and should never be handled without proper, responsible adult supervision. Using politically neutral, culturally diverse information, children are taught to: stop, don't touch, leave the area and tell an adult if they find a gun. The program is easily taught and has demonstrated effectiveness in numerous specific incidents around the country. The Eddie Eagle Program can be credited as one of the major reasons that fatal firearms accidents have been reduced to the lowest point since statistical data has been collected on this subject. With firearms in more than half of American homes, it is critical that children be given easy-to-remember instructions to follow in the event he or she finds a gun.

The program can be enhanced for the school setting by officers stressing to students that they should never carry weapons to school and that students, as responsible citizens, should get involved and notify a staff member if they know that a weapon is on campus. By teaching the program to elementary students, the tone is set early on that weapons should not be brought to school, and that they should become involved by reporting school weapons violators. Later courses such as Law Related Education can be utilized to reinforce this message in a different context. The program is available through the Community Service Programs Division of the National Rifle Association at 1-800-231-0752.

- **Gang Resistance, Education, and Training Program (G.R.E.A.T.©)** - This is another nationally recognized program designed to teach young people why they should not become involved in gang activity and how to avoid pressure to join a gang. With the rapid growth in youth gang activity, this program has proven to be effective in reducing such activity even in gang infested communities. Equally important is the use of the G.R.E.A.T. program as an effective preventive measure. Communities that do not yet have a serious gang problem can implement G.R.E.A.T. as an effective preventive measure to help keep gang activity out of the community and thus out of the schools.

Officer Earl Evans was the first of eight Bibb County Board of Education Campus Police Officers to be certified to teach the Gang Resistance Education and Training Program (G.R.E.A.T.). When the department did not have funds to purchase T-shirts for the students during the first year he taught the program, Officer Evans purchased $1,200 worth of T-shirts at his own expense. The dedication of Officer Evans and his fellow G.R.E.A.T. instructors have resulted in the department being featured as a model G.R.E.A.T. program and receiving more than $750,000 in G.R.E.A.T. grant funding.

G.R.E.A.T. is a highly structured program that can only be taught by experienced officers who have been certified as instructors through a rigorous training program. G.R.E.A.T. is primarily designed to be taught in nine sessions at the middle school level. The program does have a five-lesson component designed for elementary students but the middle school program is by far the most popular version. G.R.E.A.T. is modeled after the Drug Abuse Resistance and Education Program (D.A.R.E.) G.R.E.A.T. is designed to compliment D.A.R.E. and the two programs are most effective if both are taught in the same school system at different grade levels. G.R.E.A.T. was developed by the Phoenix, Arizona Police Department and is supported by the Bureau of Alcohol, Tobacco, and Firearms. G.R.E.A.T. program personnel can be reached at 1-800-726-7070.

- **Drug Abuse, Resistance, and Education (D.A.R.E.)** - This well known program has been taught in U.S. schools since 1983. The program was developed as a cooperative effort between the Los Angeles Police Department and the Los Angeles Unified School District. The program is designed to teach elementary school children the skills they need to resist the peer pressure of experimenting with drugs, alcohol and tobacco. While some have questioned the effectiveness of the program, many veteran officers feel that this program is a critical component in law enforcement efforts to combat drug abuse. The D.A.R.E. Program may be contacted at Post Office Box 512090, Los Angeles, California 90051-0090, http://www.dare-america.com.
- **Bicycle Safety** - Programs that include safety seminars and rodeos and focus on bike, skateboard and in-line skate safety can be an important means to save young lives. Just as the Eddie Eagle Program has educated millions of children to reduce gun accidents, these programs can help reduce this leading cause of death among children. National Safety Council Data indicates that deaths of children due to head injury from falls ranks second only to auto accidents. Officers who are assigned to mountain bike patrol are naturally suited to teach programs of this type. Often, enforcement and educational campaigns can be put into place together to yield the best results by increasing the use of helmets and other safety equipment by children.
- **Drunk Driving Accident Simulations** - Accidents resulting in death or serious injury resulting from operation of motor vehicles by young drivers under the influence of alcohol is of vital concern within the school community. The occurrence of a graduating senior who is killed in a drunk driving accident after the prom or graduation exercise is a common news story repeated across the nation each year. One effective type of preventive measure for this devastating problem is a carefully planned mock accident event. Coroner Ed Bond of the Bibb County, Georgia, Coroner's Office has developed an excellent program of this type using student actors, police, fire and other emergency personnel. Junk cars are used and emergency per-

sonnel use rescue equipment such as "the Jaws of Life" to cut open the cars in a dramatic event that drives home the finality of death due to drunk driving. This type of program usually generates a great deal of assistance and support and can make a worthwhile and lasting impression on students. For information on a truly top flight program of this type contact the Bibb County Coroner's Office at 912-749-6314.

These are obviously only a few of a wide number of quality ideas along with the formal programs available to implement them. A review of these and others provides an idea of how valuable education can be in developing an effective and high quality school police program.

Chapter 4
WEAPONS ON CAMPUS

Unfortunately, the threat of a serious weapons incident is present in every public and private school in our nation, recognizing no geographic or socioeconomic boundaries. In recent years, major shooting incidents have occurred even in small school systems around the nation. There has also been an increase in school bomb incidents. In January, 1999, a bomb detonated in a mid-western high school and injured 11 students. In another case at a high school in a small town a student was found with a book bag full of dynamite set with an armed detonator and timing device. Explosives were utilized in the horrific act of student violence at Columbine High School in Colorado in the spring of 1999.

There have been a host of other less publicized incidents involving firearms, explosives and incendiary devices in schools. It is critical that we understand the increased danger of acts similar to those of domestic terrorism in the school setting. Acts of large scale violence designed to make a statement, garner a great deal of media attention and to cause great fear have become more common in the school setting during the last several years.

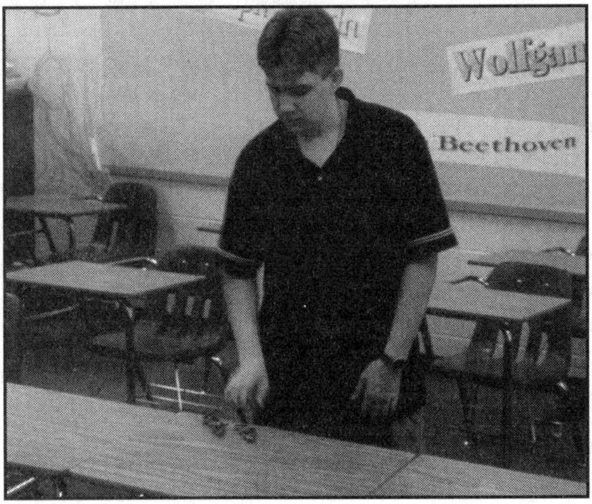

The author's son Chris Dorn beginning one of his weapons concealment demonstrations. Chris is a popular presenter and has now trained thousands of educators and public safety officials at state, national, and international conferences in fifteen states.

As Chris swiftly moves through his 43-second demonstration, guns are produced from his pockets and untucked shirt to demonstrate the importance of dress code. Fortunately for the rest of us, Chris intends to pursue a career in law enforcement after he completes college.

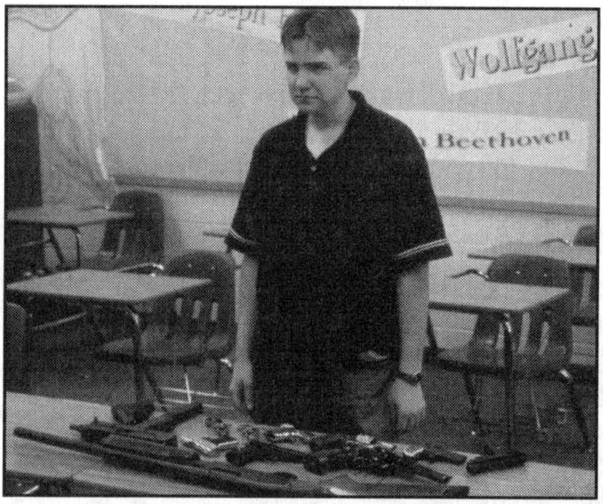

Chris has completed his demonstration which involves 12 firearms. During his record demonstration at a conference in Montana, Chris was able to conceal 66 weapons including a full size M-16 Rifle, a full length shotgun, and a wide variety of handguns, knives, brass knuckles, two machetes, two hand grenades, and a bayonet. A version of this presentation can be seen in the video "Safe Schools - Prevention, Planning, and Response" that is available free to school and public safety officials by request on official agency letterhead. Call Garrett Metal Detectors at 1.800.234.6151 to request video.

During the fall of 1998, the author participated in arresting two gang members from different private schools in separate incidents where these students attempted to bring loaded handguns into extra-curricular public school functions. One of the suspects told officers that he came to the function to find a rival gang member so he could shoot and kill him. The other suspect attempted to grab his gun from under the car seat during the arrest, indicating an apparent willingness to shoot his way out of the situation. These and many incidents of violence in the school setting indicate the necessity of concern over the problem of student weapons violations.

In example after example serious injury or death occurs in a school. Often, this happens in places where we have unfortunately grown accustomed to expect it-schools in urban centers and communities with severe crime problems. But increasingly, the words "we never thought that it could happen here" are spoken after a serious weapons assault takes place.

What is to be done? Do we turn our schools into maximum security facilities? Do our students in some extreme cases already feel imprisoned by fear because we are not doing enough? How do we find the happy medium for these two extreme versions of a prison/school?

First, we must recognize several key concepts:

- What works in one school setting may not work, be legal, be adequate or may be "overkill" in another school setting.
- No matter what measures are taken, a weapons assault can still take place, even though the danger may be reduced to the point where such an incident is extremely unlikely.
- A weapons screening program is only as effective as its weakest component. This relates directly to the quality of personnel, equipment and strategy.
- The problem of weapons on and near campuses is complex. Only a comprehensive strategy will be adequate to deal with the threat.
- Poorly designed and implemented programs usually yield poor results and may create new problems while not addressing the target problem.

Given these facts, there are a number of individual components that can serve as specific elements of an effective strategy appropriate to the needs of any specific school. These general guidelines can be tailored to fit the challenges and objectives of any setting with measures adjusted periodically to meet changing needs. Before we decide on the cure, however, it would be well for us to examine the types of weapons violators and the weapons involved.

First, we should look at weapons carried on a violator's person. This is of greatest concern because these are the weapons most often utilized when an injury or death occurs on school property. There are six types of scenarios to consider:

1. Items that are carried by the violator for reasons other than to use as a weapon. These weapons are sometimes used in the heat of the moment even though they were not necessarily brought to the campus to be used against another person. A good example of this type of item is the pocketknife that is carried routinely to school as a tool by a student who intends no harm. During an argument or fight, the pocketknife is produced and used as a weapon. This type of violation can be reduced through effective information and educational programs.

2. Items that are carried on an occasional basis as a weapon. This is typified by the student who carries a weapon to school from time to time when he or she is faced with some type of conflict. Typically, this student has been threatened or assaulted while en route to and from school rather than on school property. This type of violation can be reduced through effective school safety zone enforcement measures.

3. Items that are carried on a regular basis for use as a weapon. This is illustrated by the student who carries a gun or knife as a matter of routine for "protection" from real or imagined threats. Often prepared to use the weapon, this type of violator is the most likely to possess a disguised weapon such as a ballpoint pen or lipstick container that contains a hidden blade. Random metal detection programs, effective school safety zone enforcement measures, dress codes requiring that shirts be tucked in at the belt line and programs to prevent triggering behaviors such as bullying can be effective in dealing with this type of violator.

4. The "show and tell violator". This student brings a weapon to school to impress a peer group, typically intending no harm to anyone. An accident or injury can take place, however, during an unexpected altercation. This type of violation can be dealt with through information and educational programs, anonymous reporting systems, policies that require discipline and prosecution of each and every student that handles such weapons and an effective random metal detection program.

5. The "trader" is a student who brings a weapon to school to sell or trade for something else. The weapon carried by the "trader" will most often be a stolen firearm. This type of violator is best dealt with through aggressive prosecution on all available charges (i.e. gun possession and theft by receiving stolen property), going to the student's house and searching for other weapons and stolen goods when a gun is seized, anonymous reporting systems and random metal detection.

6. The high-risk victim violator is a student who has been shot at, beaten severely, or attacked off campus. Fearing that the violator(s) will attempt to "finish the job", this student feels compelled to go armed despite the risk of being caught and severely punished. He or she feels that "it is better to be tried by twelve than carried by six" and is more afraid of the real or perceived threat of injury or death than of going to jail. Most often, this student will be involved with gang activity and is afraid of rival gang members. This type of violation requires a great deal of intervention.

Such intervention will result from easily accessible assistance from administrators and on-site police personnel, effective school safety zone enforcement, excellent intelligence capabilities relating to off-campus gang activity, law enforcement reporting to school officials of student involvement in off campus gang-related incidents and frequent random metal detection. Aggressive use of stop-and-frisk firearms interdiction can be useful in addition to a protocol to increase all preventive search operations whenever a major gang shooting takes place in the community. This conditions gang members to expect a greater probability of being searched whenever tensions are high. In cases where this type of situation is a regular problem, comprehensive entry-point screening combined with random surprise metal detection may be needed to provide a reasonable level of safety.

7. The most dangerous weapons violator and the most challenging threat to school police programs is the "plotter". This is the determined student or non-student who plans to commit a serious weapons assault at a school. This individual normally accepts that he or she will be caught and face a lengthy prison sentence or be killed. We must make no mistake in realizing that this individual must be likened to a terrorist who is acting alone. As sad as it is, counter-terrorism strategies are required to make it likely that such attempts can be prevented or thwarted. To make matters more difficult, the worst cases of this type have typically taken place in school systems where the perceived threat level is low prior to the incident because these locations typically have few target-hardening measures in place. The problem is, of course, that it is difficult to deter a violator who is not concerned with the reality of serving a lengthy prison sentence or being killed. Comprehensive measures can attempt to cause this type of violator to shift to another target location or reduce the amount of damage inflicted by increasing the odds of neutralizing a major weapons assault. Strategies that can stand the best chance of serving these purposes include:

- Comprehensive safe school planning with well thought out critical incident management measures put into place
- Mock incident exercises
- Well armed and tactically trained on-site police personnel
- Periodic joint training exercises between school police personnel and police tactical units that take place at school sites, preferably with school staff involved for role play
- Frequent use of random metal detection
- Effective intervention by staff members whenever they become aware that a student is being repeatedly teased, bullied or ridiculed by other students
- Policies that reduce triggering behaviors
- Effective monitoring of students and non-students as they enter campuses with an emphasis on inspecting large carry items such as duffel bags, suitcases, musical instruments and the like that might contain long guns
- Use of a multi-disciplinary threat assessment team utilizing a law enforcement officer, school administrator, and a mental health professional to investigate reported threats.

Training for school personnel relating to techniques to spot persons who may be carrying or concealing a weapon concealed on their person and a practice of any such persons being patted down by armed officers can also be worthwhile. Another measure that can deter this type of violator is to require that any large item such as a science project be brought to campus by a parent. Although these measures will greatly enhance the safety of the campus, only very comprehensive measures can create an effective "hard target", and even they will not be fool proof. Such measures would include effective entry-point weapons screening with reliable access control at all school entrances, armed police personnel on duty at each weapons screening checkpoint, maintaining a sterile environment where a violator cannot enter the facility after hours to hide a weapon and later return to commit the assault, effective armed police presence in all parking areas and outside areas where students congregate.

Preventive measures that have been mentioned in this chapter will be discussed in further detail elsewhere in this text. It is apparent that a realistic threat assessment is needed before an effective overall strategy is designed because the problem cannot be properly fixed until it is defined. It is also obvious that there are no easy answers or quick fixes to the problem of weapons in schools even though careful assessment and planning can greatly reduce the threat level at any particular site or event.

What type of weapons are carried to and used on our campuses? The only limit here is the imagination and determination of the violator. Weapons seized and used in schools range from edged weapons (by far the most commonly carried and used) to impact weapons (most often found in vehicles), pepper spray and chemical agents (most often found in pocketbooks), long guns (usually found in vehicles), and handguns. Of increasing concern is the threat of explosives, incendiary devices and weapons of mass destruction such as sarin which can be made from instructions available on the internet.

The weapons can include a wide array of commercially available or homemade disguised instruments. Most of these fall into the edged-weapon category. Some commonly seen versions include pens containing knife blades, lipstick knives, belt buckle knives, umbrella knives and even pen-guns, toy guns that have been modified to fire real ammunition-or zip guns.

A quality weapons screening program must consider all types of disguised weapons to be effective. The videotape "Concealed Weapons" is available free of charge from Garrett Metal Detectors to any school system, police department or police academy that requests a copy. This video features a 15-minute presentation illustrating the wide range of weapons that are often concealed on a violator's person. The video may be requested by fax at (972) 494-1881 or e-mail at security@garrett.com.

It is important that school staff and school police officers have training in recognizing weapons, spotting indicators that a person is armed and learning how to react when a concealed weapon is discovered. There are specific patterns of behavior that frequently indicate that a person is carrying a concealed weapon. These indicators correlate to the type of weapon and the method of concealment:

- Violators carrying handguns on their person or accessible in vehicles

- Violators carrying long guns on their person
- Violators carrying knives on their person
- Violators carrying weapons in book bags
- Violators carrying weapons in purses
- Violators keeping weapons in school lockers
- Violators hiding weapons in public areas
- Violators hiding other weapons in vehicles.

Let's examine each of these scenarios individually before we cover weapons screening techniques in the next chapter.

Handguns carried on the person or in vehicles - The vast majority of individuals who carry a handgun illegally on their person do not utilize a holster. Most police officers and citizens who legally carry a handgun in compliance with the laws of their state, use a holster or other carrying device. Most criminals however, are not very concerned with safety and simply stick the gun in their pocket, or tuck it in their waistband.

Frequently, this type of carry makes the violator easy to spot. On many occasions a handgun has been recovered because the outline of the gun was discernible to the trained observer. In some cases, part of the gun itself could be seen protruding from clothing. In many cases, a prominent bulge will be seen, even if the outline of the gun is not visible. A handgun carried in a jacket or coat pocket will also often cause the garment to sag noticeably on the side where the weapon is hidden. Also, a gun in a jacket pocket will often cause the pocket to swing unnaturally when the wearer walks or runs.

Handguns are sometimes carried in "belly bags" which have become popular in recent years. In fact, there are numerous bags on the market which contain hidden holsters to secure a handgun. Although there are a wide variety of holsters on the market, several particular styles bear mentioning. First, the ankle or leg holster can be used to hide a weapon under a boot or pant leg. Often, a person wearing such a holster will walk with an abnormal gait. Also, a bulge may often be seen in the pant leg or on the side of the boot. Next, a variety of shoulder holsters help to conceal a weapon under the arm. Again, the weight of the weapon may cause an unnatural swing when the wearer walks quickly or runs. Frequently, a bulge can be seen behind the armpit of the wearer. Shoulder holsters as some hip holsters may cause the wearer to have to pull one side of a jacket down to prevent the bottom of the holster from being visible.

Individuals who unlawfully carry a handgun often give off visual clues caused by nervousness., The discomfort caused by the presence of the weapon or an automatic reaction to a weapon that may or has shifted may often be a visual indicator. The gun violator will often show signs of nervousness by trying to ease away upon the approach of a police officer. The violator will also often turn his or her body away from an approaching police officer. Sometimes one person in a group has been designated to carry a gun; that individual will often be standing near the group, but separate from them. Often, a student who is carrying a weapon will run when approached by school officials or police. This often enables the violator to dispose of the weapon. This is one reason that students who run from school police should be charged with the appropriate code violation whenever possible; otherwise, any student with drugs or a weapon would be foolish not to flee when confronted.

In the fall of 1997, the Bibb County Board of Education Campus Police Department was awarded a technical assistance grant from the United States Bureau of Justice for the investigation of firearms crimes and for interdiction of illegal firearms. The grant was administered by the International Association of Chiefs of Police. This grant proved to be very effective in increasing the department's capabilities to interdict illegal firearms. During the first 30 days after technical assistance was received, the department recovered seven handguns from violators. These arrests included:

- One public middle school student who was arrested after a campus police officer observed that his jacket was swinging in an unnatural manner as he walked down the hallway in the morning. The officer recovered an unloaded .25 caliber semi-automatic pistol from the student, a known gang member.
- A convicted felon who was stopped for a loud music violation on a city street next to a public high school. A loaded high capacity nine millimeter handgun was found under the driver's car seat during a consent search. The suspect who is a gang member, had 26 previous felony arrests.
- A student, from a private school, who was riding in a car with three other gang members in the parking lot of a stadium during a football game. The 14 year-old youth was arrested after a consent search revealed a loaded .380 caliber semi-automatic handgun was tucked in his front right waistband (the suspect is right-handed). The suspect stated that he was looking for a rival gang member who had shot at him on an earlier occasion. He stated that he thought that the suspect might be at the game and that he intended to "find him, shoot him and kill him".
- Another student, from a parochial school, was riding in a car stopped for cruising in the same parking lot during a football game. The driver gave consent for the vehicle to be searched. The student then attempted to grab a gun from under the seat when he learned that the car would be checked for weapons. The student's actions and the circumstances indicated that he was willing to attempt to shoot it out with the three officers at the scene. This student initially denied that the gun belonged to him until officers found his full name inscribed on the pistol's magazine. The 16-year-old

finally admitted ownership of the gun and said that he had been carrying it regularly for three months because other young people had repeatedly assaulted him for wearing gang colors.

- An out-of-state, non-custodial parent was arrested after he appeared at an elementary school requesting permission to see his child. When school officials checked the child's emergency contact card, they found notations that indicated that the father had threatened to kill the child and his mother. Officers also learned that the suspect had three prior arrests for domestic violence. The suspect was found parked across the street waiting for school dismissal after he was denied access to the child, and when asked by responding officers if he had a gun in the vehicle, he admitted that he had a .40-caliber semi-automatic in the vehicle. Several other weapons were found in the suspect's vehicle.

While only one of the seven arrests involved a public school student, it should be readily apparent that any of these situations could have resulted in a very significant problem if alert officers had not found the weapon before it could be used to commit a crime on or next to a campus. As part of the technical assistance grant, Sergeant G.G. Neill of the Washington D.C. Metropolitan Police Department was flown to Macon where he conducted a training session for the department and spent time in the field with the Campus Police Department Special Operations Unit. Sergeant Neill was able to provide considerable input in addition to the information already available to the department.

The gun violator will often give observable cues such as the "security check". During the author's visit to IACP headquarters in Alexandria, Virginia, Sergeant Mark Cowley from the Baltimore County Police Department explained this concept to the author. A security check is a frequently observable and strong indicator that a person is carrying a handgun. A person carrying a handgun that is tucked in the waistband or in a jacket pocket will periodically check the weapon. This "security check" may be a simple pat or a more noticeable effort to reposition the weapon. The security check will often be performed when the violator stands up after squatting, sitting or when getting in or out of a car. Sergeant Cowley has also observed that on many occasions, a gun violator will not swing his arm as much as normal or at all on the side of the body where the gun is concealed. The author has observed this same behavior with knife and drug violators, and has also observed that a gun violator will sometimes swing the arm unnaturally, extending the elbow outward to avoid hitting the butt of the gun.

The author recommends approaching a potential gun violator from the gun-side rear whenever possible. The gun side can sometimes be determined with probability by looking at the suspect's watch. A right handed person normally wears a wristwatch on the left hand with the watch facing outward. A left-handed person often wears the watch on the right hand. Typically, most people often take their first step with their leg on the weak side. By approaching from the strong side rear, the officer will have much more warning if the suspect produces a gun before turning to shoot at the officer. If the offender has to pull the gun and turn to his gun side, he or she will have to expose the gun as it is swung around to bear on the officer and, thus, must shoot from an awkward position. If the officer is directly behind the suspect or worse, to the weak side rear, the violator can produce the weapon and turn to fire with little warning for the officer who may not see the weapon until it is pointed in the officer's direction.

If the strong side cannot be determined, the officer should approach from the right rear as most people are right handed. Of course, available cover should be an overriding factor in this type of stop. If a solid mass that will stop bullets is available, it should be utilized whenever it is appropriate regardless of which side it is on in relation to the suspect. As always, it is strongly advisable whenever possible that more than one officer approach suspects who are potentially armed. It must be remembered that even though proper approach in the above manner can provide three to four times more reaction time for the officer, a suspect can still often turn and shoot with great speed.

Even though there is no legal ownership of handguns or handgun ammunition in the District of Columbia, the district has an enormous problem with handgun crimes. Sergeant Neill works in an environment where handgun violators are very common. His assignments for several years have been focused on street-level firearms interdiction, where he has spent considerable time honing his skills at illegal gun recovery. Sergeant Neill has recovered more than 700 illegal firearms using the techniques described in this chapter. Working in an environment that is rich with gun violators, he has developed exceptional skills at locating, identifying and prosecuting the gun violator. Sergeant Neill like Sergeant Cowley relies heavily on the observations made by Detective Robert Galligher of the New York City Police Department. Detective Galligher retired after making more than 1,200 arrests on gun charges during his 30 years on the force. He is the most highly decorated detective in the history of the New York City Police Department. Again, working in an environment where illegal handguns are prolific, this officer was able to note a number of characteristics that are exhibited by the armed criminal. It is crucial to understand that there is no "profile" of the gun violator. There are instead, certain specific actions that can frequently be observed being taken by the gun violator. Being an astute observer of human behavior, Detective Galligher organized his observations to help other officers learn to spot gun violators. Very few police officers work in settings where illegal firearms are so commonplace, and would probably never make many of these observations without exposure to training in Detective Galligher's observations.

It is important to realize that one factor alone may not provide legal basis for a frisk for weapons. Often, an observant officer can combine this factors with other observations to develop legal justification. Also, officers may ask for and obtain lawful consent from a suspect to search when one or more indicators are present.

Sergeant Neill has also very successfully utilized the "rainy day rule". Sergeant Neill has made numerous arrests by observing pedestrians during periods of heavy rainfall. He has found that a person who tries to run with a gun tucked in a waistband, in a shoulder holster, or in a jacket pocket will normally hold the gun when they run. The violator has to hold the gun in this manner because of the weapon's weight. A gun in the waistband will often fall out if it is not held in place during periods of physical exertion. A gun in a shoulder holster or jacket pocket will often swing uncomfortably if the violator does not hold it. As it is normal to see people run when it rains, this is an excellent time to spot gun violators. In a school setting, it could be very useful for an officer to observe students as they run from the parking lot or from the bus line to the building.

This information is also valuable to officers engaged in a foot pursuit. The author has on many occasions been able to tell that a suspect was armed because the suspect held one hand on a gun tucked in a waistband or pocket during a chase.

Another useful cue for police officers who stop an individual who is carrying a gun unlawfully is the tendency for the violator to keep his gun side away from the officer. The violator may do this from fear that the officer will see the weapon, because they plan to use it to attack the officer or simply as a natural "gut reaction" that so often results when a suspect is approached by an officer.

Officers should also be alert for protrusions in the clothing that appear and disappear as the violator rotates the torso. Often, a small protrusion will appear as the clothing binds on the sharp corners of the weapon. Even very small guns may cause this telltale sign that something may be amiss.

Another visual cue is the tendency for the gun violator to take a shorter leg stride on the side of the body where the weapon is concealed. This is even more common with large weapons such as sawed-off rifles and shotguns that are sometimes hidden in the waistband and extend down the pant leg. In this instance, the stride may be changed so much that a limp is observed.

Other cues may come from the suspect's clothing. Officers may notice the sag of a jacket and the unnatural swing of a jacket pocket from the weight of a gun. Detective Galligher also points out that the fabric on the non-gun side of the jacket will tend to cling to the front of the pants, while the fabric on the side where the gun is carried often swings clear of the hip when the suspect is walking. Another trend is the overly baggy clothing worn by many young people. Such clothing can make it particularly difficult for a teacher or officer to observe a student who may be carrying a firearm. In Bibb County, students are required to wear appropriately sized pants worn at the waistline.

Of course, any officer should be alert to people who wear clothing that is noticeably heavy for the current weather. Often, a suspect will wear a jacket or coat to hide a weapon even during warm weather. Another cue may be a jacket that is obviously mismatched to the rest of the person's clothing. This may result from a suspect using the mismatching jacket solely to hide a weapon. Another cue may be seen when an individual wears only one glove during very cold weather. This allows the violator to use a gun with the ungloved hand. Thick gloves may make it difficult or even impossible to handle a gun, particularly a small handgun.

Detective Galligher also noted that when a gun is stored in a carry bag, it may lean heavily in a particular direction and tilt down in an unnatural manner.

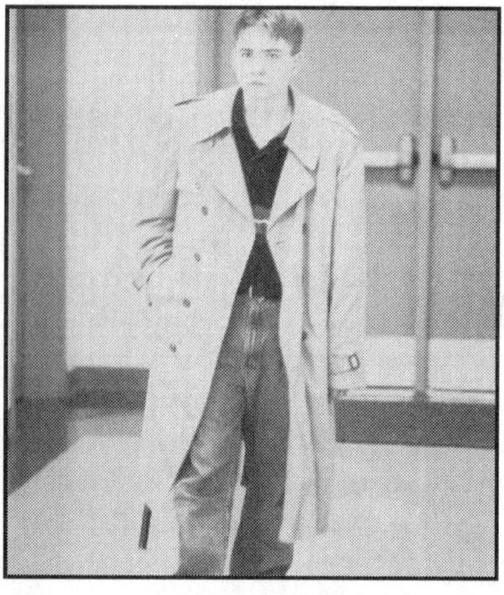

While it may seem obvious that this individual is concealing a long gun, such violators have walked into American schools undetected on a number of occasions. Note also the manner in which the jacket lays unnaturally on the gun side and the position of the violator's elbow and shoulder. This would make detection possible even if the muzzle of the weapon were not visible.

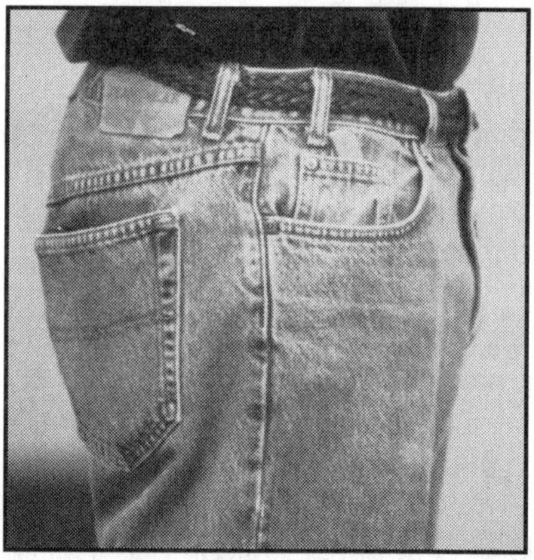

A pocket knife can be seen sticking out of the violator's pocket. While most visual screening techniques are geared towards firearms, knives, impact weapons and explosives devices, many other types of weapons have been successfully detected with visual screening.

In some cases, the outline of the gun may be seen. The author once observed the end of a shotgun barrel protruding from a shoulder bag during a nighttime stop of a suspicious person. Often, a carry bag will be left unzipped to allow quick access to a firearm.

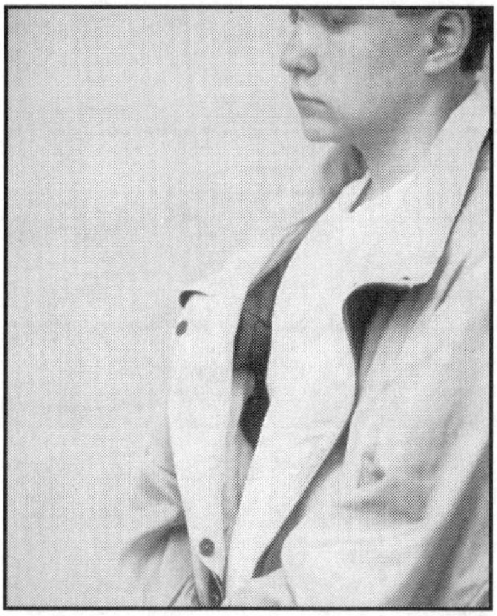

In some cases, a part of a handgun in a shoulder holster or a long gun which is being concealed under a jacket or coat can be observed.

Often, a concealed firearm will result in an observable bulge that may be located in front of or behind the armpit of the violator, in the waistband area, on the torso or in the ankle area as the violator moves. It should be noted that the arrest of the Oklahoma City bomber was the result of an Oklahoma Highway Patrol Officer who observed a bulge in the suspect's clothing during a traffic stop. Upon frisking the suspect, the officer recovered a handgun from the suspect and took him into custody.

In many instances, the outline of a weapon can be seen. In this case, the outline of a small handgun can be noted by the keen observer. In many cases, knives, metal knuckles and other weapons have been detected. Other contraband such as pocket pagers and cigarette lighters have been detected.

When even a small handgun is carried in a jacket or coat pocket, the garment will sag on that side. The weapon will also frequently cause the garment to hang in a taunt fashion and the weight of the weapon may cause the pocket area of the garment to swing in an unnatural fashion.

This photo shows the need to be alert to innovative violators. A tip was received by a Bibb County Campus Police Officer that one of two juvenile students had a gun at school. When they were searched, one of the students had several rounds of .45 caliber ammunition in his pocket. Officers obtained consent to search both students bedrooms from their parents and found this tennis racket bag under one student's mattress.

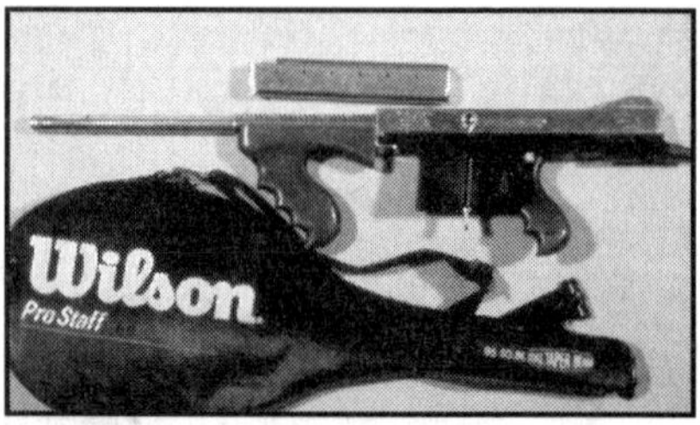

Inside the bag was this .45 caliber 25 shot semi-automatic carbine. The gun and the bag had been extensively altered so the students (who were gang members) could carry the weapon with less chance of being detected. This case also demonstrates the need for the use of the home search technique to recover illegal firearms and explosives before they are brought to school.

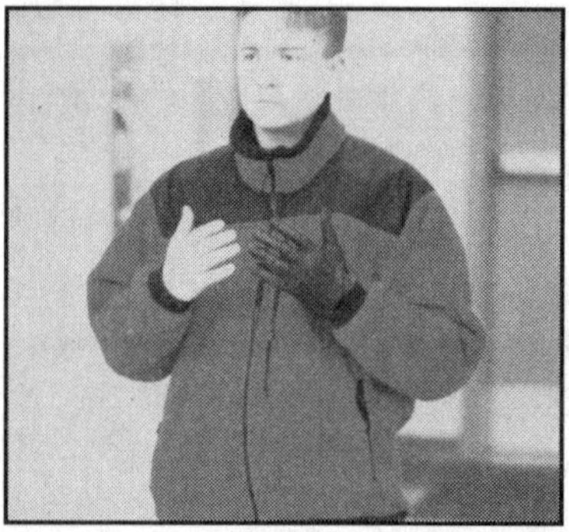

In cold regions a gun violator may wear a glove only on the weak hand which allows the violator to draw and operate the weapon without interference from a glove.

Sergeant Neill credits a high percentage of his gun recoveries to vehicle stops. Nationally, traffic stops account for the highest percentage of arrests for firearms-related crimes. Sergeant Neill uses a very direct approach to seek consent to search vehicles that he thinks may contain an illegal firearm. After making contact with the driver, Sergeant Neill asks the driver, "Do you have any drugs or guns in the car?". In some cases, the driver will admit that drugs or illegal guns are hidden in the car. Normally, the driver will deny that any contraband is in the car, and Sergeant Neill then asks the driver, "Do you mind if we search the car for drugs and guns?". Sergeant Neill rode on patrol with three officers from the Bibb County Board of Education Campus Police Department's Special Operations Unit for a four-hour period. During this time period, the officers made a number of traffic stops for moving violations. Sergeant Neill then attempted to obtain consent to search the vehicle from the driver and in every instance except one, he was easily able to do so.

Most of the seven weapons arrests in the 30-day period described earlier were made in this fashion. These types of stops illustrate the great utility of traffic enforcement. Such enforcement can be very effective at intercepting illegal firearms before they come onto school property, and in cleaning up school safety zones. Aggressive enforcement can be a strong deterrent for the criminal who might otherwise frequent areas close to a school and pose a threat to students as they attempt to walk or drive through the area in the mornings and afternoons.

Traffic checkpoints when combined with traffic enforcement focused on the interdiction of illegal firearms and drugs is one of the most effective means to prevent drive-by shootings at and near schools.

One very important behavior to look for is the "head turn and shoulder dip" that is sometimes seen on traffic stops. In this instance, the violator has drugs or a weapon on his person or in plain view in the car. When a police officer is approaching the suspect's vehicle, he turns slightly to watch the officer in a rear view mirror while attempting to hide the contraband under the seat of the car. An observant officer will note the slight turn of the suspect's head and the dip of one shoulder as the suspect tries to discreetly hide the contraband. More importantly, this behavior can be a warning that the suspect has reached down to get a gun.

Sergeant Neill also emphasizes the critical importance of noting the positioning of the car's occupants at the time of the stop. This can become an important issue in successful prosecution when contraband is found hidden in the car during a search. In many instances, there may be a number of individuals in the car and the officer may need to demonstrate in court the access to the evidence by a particular person.

One tactic utilized by Sergeant Neill is to have occupants step out the car one at a time while a second officer watches. Each occupant is searched while standing in the "V" of the open car door. By positioning each occupant between the car, the open door, and the officer, the need for foot pursuits can be greatly reduced. After each passenger is searched, the vehicle is checked for weapons and drugs. It is critical that occupants be carefully observed during this process so the searching officer is not attacked with a weapon, or that evidence cannot be passed among passengers.

It is important to watch for occupants who may try to drop drugs or weapons on the ground as they step out of the vehicle. Often, these suspects will try to kick the evidence discretely under the car so it cannot be discovered by officers. It is a good idea to routinely look under a suspect's car after all occupants have been removed in case any of the occupants has managed to dump evidence in this manner. Commander Steve Meadows from the Bibb School Police Special Operations Unit made an arrest after he observed an occupant drop a small packet through a hole in the floorboard of the vehicle during a stop. The packet was recovered from the pavement under the car and found to be counterfeit cocaine.

Bibb County Campus Police Officers issued more than 2,000 traffic tickets and made over 300 arrests as a result of traffic stops in a 90 day period. These efforts stopped a series of six gang related shootings that had taken place on city streets adjacent to public schools during the first three months of the school year. This countermeasure proved to be extremely effective when an explosion of gang activity occurred in Macon, Georgia.

It is important that officers look methodically but quickly for drugs and firearms during vehicle searches. Officers who tend to detain motorists for lengthy periods during consent searches may find that consent is withdrawn by many drivers or may find it increasingly difficult to obtain consent if they gain a reputation in the patrol area for detaining people for unreasonably long periods of time. With practice, a car can be searched carefully in just a few minutes on most occasions. Of course there are exceptions, but long searches should not be the norm.

Officers must be alert to hidden compartments either of factory design or from modifications. Officers must be alert to screws and bolts that show signs of being removed repeatedly. They should also look in cavities such as open areas in sunroofs and under removable items such as ashtrays and containers in center consoles. Often, particular vehicle hiding areas become prevalent in a geographical region because violators share their expertise with each other. While most firearms will be recovered from under a car seat, in a console, door pouch or a glove box, there are a number of other potential locations where a gun may be hidden. They can often be found by using a flashlight to look carefully under the dash and under seats. Offenders will sometimes rig holsters or other devices to hide the weapon out of sight but within easy reach.

According to a study by the International Association of Chiefs of Police, the traffic stop is the police activity that is used to recover more illegal firearms than any other method.

Long guns carried on the person - In the vast majority of cases, long guns seized on school property will be found in vehicles. In most cases when an individual enters a school with a long gun that is concealed on the person, the violator intends to threaten or assault someone with the weapon. This is because a long gun is normally uncomfortable to carry concealed for any length of time and it is even more difficult for the violator to remain undetected for long periods of time. This is true even with sawed-off rifles and shotguns. Violators may also attempt to conceal long guns under a jacket or coat. The violator may hold the weapon in place by grasping it through a pocket. Another method that is utilized is to rig the weapon with a strap or sling to hang the weapon over a shoulder. The long gun violator may also utilize a carry bag, book bag, duffel bag or other such case to hide the weapon. The weapon can also be hidden by draping a jacket over the weapon or by wrapping the weapon in a jacket or blanket.

This type of violator can often be spotted by an alert officer who notices someone walking in a stiff manner, with an unusual gait or an unnatural limp. If a long gun is tucked down the pants, the weapon may produce a noticeable bulge in the pant leg or on the back. The bulge may appear and disappear again as the violator walks. This type of violator will normally be unable to swing the arm on their gun side. Again, the violator may also exhibit clues by performing a "security check" which will be more noticeable than with a handgun because of the increased weight and size of the weapon. The long gun violator will not normally be capable of sitting in a vehicle or chair because of the length of the weapon. This can be helpful because the violator who rides to school in a car may have to risk being spotted when they get out of the car and try to hide the weapon before entering a school building. This also makes it unusual for a student to ride a bus to school while hiding a weapon in this manner. With short sawed-off rifles and shotguns, the violator may be able to sit with the weapon concealed.

Knives carried on the person - Knives carried on the person constitute the most common type of weapon and method of carry that results in a school weapons assault. Knives are easy for students to acquire and conceal. They cannot be traced and, most importantly, are not treated as a serious threat by many educators, police officers and courts. Many school systems do not deal firmly with knife possession cases unless there is an assault or threat made by the violator. Until recently, one large Southeastern school system even allowed students to carry knives if the blade was less than two inches long.

Knives carried to school concealed on the person will normally be found hidden in one of the violator's pockets (with male violators). Small coin pockets are a favorite and often overlooked concealment location. Often, small knives will be found attached to a key ring. Small pocket knives and box cutters are the types of edged weapons most frequently carried and the type most frequently used in school knife assaults. Again, this results from the generally mild discipline received when these types of weapons are confiscated. Students also frequently switch to knives to attempt to avoid detection when screening programs are implemented or to avoid prosecution if they are caught. While small knives are less dangerous than big ones, they still pose a serious threat.

One important difference between edged-weapons assaults and firearms assaults is the fact that people are often victimized with an edged weapon without bystanders being initially aware that an assault has taken place. In fact, in many cases, the victim is not even initially aware that he or she has been cut. The author has personally experienced this, having been cut on the leg with a razor and not realizing it until someone else noticed the damage to clothing and blood that was present. Often, knife wounds (and in some cases, gunshot wounds) do not result in noticeable pain. This is even more likely to occur when the victim is in an excited state because they have been engaged in a fist-fight with the assailant. Another phenomenon is the penetration into the body of the victim to a depth greater than the length of the blade. It is very common for a stab wound to be much deeper than the blade length because the tissue compresses when the victim is stabbed directly with force. This is another reason why a one- or two-inch knife should be taken seriously in a school setting.

There are numerous cases in which box cutters and very small knives have been utilized by violators to commit serious and even fatal school assaults. The author was recently consulted about a stabbing in Virginia where a high school student was stabbed in the chest with a pocketknife having a blade length less than one inch. The victim nearly died after his lung collapsed. The author is also familiar with a case in a Georgia school where a female student died after being stabbed with a two-inch pocketknife in the chest at an urban school. It is important that these items be defined, regulated and treated as dangerous weapons when on school property.

While pockets are the number one location for edged weapons concealed on the person, they may be found in other places such as in socks, shoes, hats, taped to or strapped to the extremities or on the torso. Knives may also be concealed in rolled pants cuffs by enterprising violators. Female violators sometimes conceal edged weapons in their hair, most often razor blades. Violators of both sexes sometimes conceal razor blades in their mouths.

While encountered less frequently, edged weapons disguised to look like innocent or novelty items are sometimes found in the school setting. These weapons may be concealed in jewelry, bracelets, ballpoint pens, lipstick containers, kubatons, umbrellas, walking canes, belt buckles, combs, hairbrushes, coins and similar items. A wide variety of disguised weapons are readily available in retail outlets, by mail order and on the internet. Officers working in the school setting must always be alert to such disguised weapons.

When dealing with persons who may be armed with an edged weapon, officers must be aware of the concept of "palming". Knife offenders will often hold a knife concealed in the palm of the hand by using the thumb or two fingers. This method allows the assailant to get close to the intended victim without warning. An edged-weapon assault normally must take place with the assailant being within reaching distance of the victim. While there are people who can effectively throw a knife, they are rare and this form of knife attack is not common. Officers and school staff should also be familiar with the concept of palming because a student who is handling a knife may palm the weapon quickly when a police officer or educator is in the area.

The knife violator may fail to swing the arm naturally on the side of the body where the knife is concealed or especially, when palming the weapon. Also, as with the gun violator, the knife violator will frequently turn the body so the side of the body where the weapon is concealed is facing away from the officer.

Weapons concealed in book bags - If entry-point screening or surprise random searches are not used, book bags will be a very common place for students to conceal guns, knives, impact weapons, alcohol, drugs and explosives. Requiring see-through book bags is not effective as students will simply hide the weapon in a container or clothing within the book bag. While some would consider clear book bags to be a solution, the author frequently gives a demonstration to show how more than three dozen weapons can be concealed in such a bag. Another means to address this problem is to require all students to store book bags in their lockers during the school day.

Some school systems have had success with totally banning book bags altogether. This approach can cause resentment because students and parents may see this measure as punishing all students for the misdeeds of a few. Most students and parents will prefer entry-point or random screening of book bags to the elimination of book bags. It is important to remember that students have been known to hollow out compartments in textbooks to create a hiding place for guns, knives and other contraband. If book bags are banned, countermeasures to deal with weapons hidden in books should be used. These can be random book inspections or the more costly measure of issuing two sets of books to each student with one set kept in the home and one set kept in the classroom. Yet another option is to issue each student a textbook to be kept at home and each teacher a master set of textbooks to be used for all students taking the same class taught by that teacher.

Weapons concealed in purses - The vast majority of weapons confiscated from female students are found in purses. With female violators, edged weapons and irritant sprays usually account for the bulk of weapons seized although other types are also encountered. Female students may also be found with electric stun guns and kubatons, which are often sold as alternatives for female self defense. Female students sometimes carry firearms for personal use, but in many cases, they are transporting the weapon for a male student. Gang members in particular, are very prone to having a female student carry a gun for a male associate. All students caught with weapons should be interviewed regarding where and why the weapons were obtained. If a female student is found with a gun, it is important that officers be particularly alert to the possibility that the weapon is being carried for another student.

In some cases, a female student who is caught with a weapon will have been given the weapon by a parent for self defense while going to and from a part time job. On some occasions, it may be found that a parent has even provided the weapon for protection while traveling to and from school or even for protection while at school.

There are purses that are manufactured with a concealed holster built into the purse to enable a woman to have safe and ready access to a handgun. These special purses are normally very expensive and are not commonly seen in most areas of the country. The author is not familiar with any purses of this type being seized at any school, but school officers should be aware of their existence. It is always possible that a purse of this type containing a handgun could be stolen and end up in a school setting.

Weapons stored in student lockers - Student lockers are one of the most common hiding places for weapons and other contraband in a school setting. Fortunately, lockers are also the easiest to deal with in an effective manner. If random locker inspections are not used, lockers will be another common location for students to hide weapons. Empty lockers left unlocked are also a favorite place for students to hide weapons.

Weapons concealed in public areas - One of the most difficult hiding places to address is the public area. Students may hide weapons in bathroom ceilings, inside the openings under sinks, under shrubbery, buried in the ground, behind water fountains, behind vending machines, behind loose bricks or any other location. Dealing effectively with public or common access areas requires more physical effort than most people might realize. It is even more difficult to apprehend and prosecute the student who hides a weapon in this manner.

Other weapons hidden in cars - We have already addressed handguns concealed in vehicles in an earlier section. At this point we shall discuss how vehicles are often used to hide a variety of other weapons. Long guns on school property are found in student vehicles more frequently than in any other location. In the vast majority of these cases, the student has left the weapon in the vehicle after target shooting or hunting. These students typically do not view the firearm as a weapon and do not intend to use the weapon to threaten or harm anyone. In the same manner, knives found in student vehicles are most often thought of by the student as a tool that they utilize while hunting, fishing, or at a part-time job.

Long guns will most often be found behind the seat of a pick up truck or in the trunk of a car. In rural areas, students are sometimes allowed to have long guns and they may be observed in gun racks in plain view. Similarly, many schools allow students to keep hunting and fishing knives in vehicles. While the vast majority of these students will probably never attempt to harm anyone with these items, allowing them on school property may pose serious liability problems for the school if an assault takes place.

Knives may be found in a wide variety of locations within a vehicle. Officers may find knives in door pockets, hidden behind sun visors, in glove boxes, or consoles, under car seats, hidden in a headrest, wedged between seats and consoles, hanging from turn signal levers, concealed under floor mats, rigged under the dashboard and in a variety of other locations. Door pockets, consoles, sun visors, and glove boxes are normally the most prevalent hiding places for edged weapons in a vehicle.

Impact weapons are also most commonly found in student vehicles because of their size and weight. Clubs, axe handles, baseball bats, sections of pipe, nightsticks, nun-chuckas and large sticks are often kept in cars as a weapon. Often, violators will write things on the item which help to show its purpose as a weapon. Tape for a grip, hand straps or added spikes or weights may also help to demonstrate the intended use of the item. In many school systems, baseball bats are not allowed on campus for students who do not play on the school baseball team. The Georgia legislature has included baseball bats in the felony weapons statute dealing with weapons on school property after cases occurred where bats have been used to commit violent assaults. As with most school policies, the state statute defines when a student may legally be in possession of a baseball bat on school property.

Impact weapons are most commonly found next to the seat by the door, under the seat, behind the seat of a pick-up truck or in the trunk of the vehicle. Often, the student will admit that the item is carried "for protection" if they are asked why they have it in the vehicle. This can be crucial to successful prosecution for a weapons charge in many instances.

The variety of weapons and the methods in which they are concealed cover a wide range of possibilities. These diverse variables are clear proof that a comprehensive approach to reducing school weapons violations is so important. A holistic approach that addresses each type of weapon and method of concealment is needed to avoid having violators simply bypass less comprehensive measures that leave easy-to-exploit gaps. Each of the threats that has been described can be reduced by specific countermeasures. In the next chapter, we shall see how each of these threats can be addressed and what countermeasures are available to do so.

Chapter 5

EFFECTIVE SCHOOL WEAPONS SCREENING

This chapter is designed to explain various preventive weapons screening concepts from an operational standpoint. Each of these techniques has proven to be effective in specific school environments where the targeted weapons were a problem. This chapter is *not* designed to be a guide to the technical aspects of metal detection. A detailed explanation of the technical side of effective school metal detection, can be obtained from *School Safety Using Metal Detectors*, which is available at no cost from Garrett Metal Detectors. This publication is the most comprehensive treatment of the topic that is available today and covers the important considerations for the use of metal detectors in schools.

In any form of school weapons screening, how it is done is just as important as what is done. Clear communications of policies, methods that will be utilized, and polite and courteous conduct by personnel are critical. All actions must be taken in a positive manner that underscores that these measures are taken for the safety of students and school employees. It is important that students, parents, staff and school police officers all understand that weapons screening is focused on deterring potential violators rather than reacting after a violation takes place. It is important that every effort be made to get students and parents to "buy into" the screening strategy. If it is done correctly, this support along with a positive public relations factor will be present. Being reasonable is not only required for these measures to be legal, it is also necessary for them to be effective.

We will explore 10 concepts of specific individual strategies that have been used to counter the specific threats outlined in the last chapter. The following concepts will be covered:

1. Perimeter screening - surveillance, stop and frisk tactics, and traffic enforcement
2. Vehicle checkpoints on campus
3. Plain-view weapons screening of vehicles
4. Canine checks of vehicles in parking areas
5. Ground area sweeps using metal detectors and canines

6. Entry point weapons screening
7. Random surprise classroom/school bus checks
8. Locker checks
9. Inside public area checks
10. Searches based on tips or observation of suspicious activity

Consider these concepts as menu items to chose from to counter specific threats. Each situation is different, and while one school may have a critical need for a specific tactic, another school may not need this method at all. The same is true with the frequency of use of individual actions. A particular type of threat may require more frequent use of a preventive tactic at one location than another.

1. Perimeter screening

The goal is to deter potential violators from ever bringing a weapon to school. When a breech takes place, the next best scenario is to confiscate the weapon before it makes it to the campus. Another concern is that students are much more prone to carry weapons to school if the neighborhood around the school is unsafe for students to walk through.

In many settings this may not be a significant problem; at some schools it is the biggest problem. Interviews with students who have been caught with a weapon at school, particularly a gun, often reveal that the student feels safe at school, but has carried the weapon to school for fear of being attacked while en route. Efforts to clean up the area around the school will therefore help to reduce weapons violations on school property. Strictly enforcing criminal and traffic codes in areas surrounding schools can be very useful in displacing criminal activity to areas away from the school. A number of state legislatures have created "school safety zones" and "drug free school zones" to enable law enforcement officers to create safer buffer zones around schools.

Several tactics can be very useful in reducing crime in school safety zones. One effective measure is to conduct motor vehicle license checks on streets around school campuses. Gang members, drug dealers and other career criminals who pose a threat to students are often habitual violators of motor vehicle laws. These individuals often have no driver's license, no insurance coverage, outstanding arrest warrants, suspended licenses, equipment violations or are driving while under the influence of drugs or alcohol. Often, these individuals will be arrested at police license checkpoints.

In many cases illegal weapons and or drugs will be found during searches incident to arrest at license checkpoints. Additionally, many traffic violations found at these checkpoints require that the violator's car be impounded and contraband is found when officers inventory vehicles prior to their being towed. Alert officers working at these checkpoints will often develop legal justification to search a vehicle, or more commonly, obtain legal consent to search occupants and vehicles.

In high crime areas, checkpoints are typically very productive in terms of arrests and citations as well as weapon and drug seizures. In many jurisdictions, police checkpoint inspections are most often conducted after dark, and the school community does not benefit. If these operations are done frequently enough during school hours, word spreads quickly among the criminal element that school zones should be avoided if you are a fugitive, driving on a suspended license, driving while intoxicated or have contraband in your car. Criminals who regularly disregard serious laws concerning drugs, weapons and acts of violence are even more prone to disregard traffic laws and are very vulnerable to police license checks. These individuals usually try to avoid areas that are frequently used by police to conduct checkpoints for this reason.

This tactic when combined with heavy traffic enforcement for moving violations also greatly reduces the risk of having a drive-by shooting on streets near a school. Often, gang members who plan to commit a drive-by shooting must "hunt" the intended victim(s) by driving around areas frequented by the victim. Gang members will sometimes try to find the intended victim by cruising around the victim's school in the morning when students are arriving, during class changes and lunch periods when students are visible on campus, and at the end of the school day when students are leaving school. License checkpoints and heavy traffic enforcement are proven ways to thin out heavy cruising around schools. This prevents a carload of gang members from "blending into" the crowd while they stalk their intended victim(s).

Typically, license checkpoints are used first. They tend to dramatically reduce the number of "undesirable" individuals who are prone to cruising in school zones. Often, parents and students are stuck in traffic each day trying to enter and leave school because of such cruising. After a series of carefully planned license checkpoints in key areas, the number of cars in the area usually drops as many of these individuals begin to get the message that cruising in these areas can result in tickets, towing of cars and arrests. Once this "thinning out" is accomplished, more reliance on moving violation enforcement can be utilized to deal with those hard-core and hard-headed criminals that still cruise in school zones.

Initially, parents and students will be inconvenienced when license checkpoints are conducted. Every effort should be made to point out to the public that even though they may experience delays, time will be saved later when they can enter and leave the campus each day in reduced traffic. Efforts should also be made to help people understand that the law does not allow police officers to pick and choose which motorists will be checked at these checkpoints if arrests are to be upheld in court. If the public relations efforts are performed properly, most citizens and students will be supportive.

Another type of checkpoint that can be conducted in some jurisdictions permits officers to select specific individuals to be stopped and searched. This type of checkpoint is conducted using an "area warrant." In jurisdictions where it is allowed by the courts an area warrant allows officers to stop motorists and pedestrians who are identified by officers as known gang members or drug dealers. The officers must be able to provide precise data showing that the location selected for the area warrant is a high crime area during time periods like those specified in the area warrant. This type of warrant allows officers to stop and search all pedestrians, motorists and vehicles that the officers can show by "articulable suspicion" to be involved with drug or gang activity during a specific time frame and at a specific place. An example would be "from 3:00 P.M. to 6:00 P.M. at the 1600 block of Jones St".

Although this type of warrant has been extensively used in some areas of the country, many police officers, prosecutors and judges have never even heard of the concept. This type of warrant will not be upheld by courts unless a severe crime problem can be demonstrated in the targeted area. While it may require a good deal of research to use this type of warrant for the first time in a given area, it can be a powerful tool to deal directly with the problem element without inconveniencing the general public.

Moving violation enforcement is one of the most underutilized and most effective means to clean up a crime stricken area. First, traffic tickets can "nickel-and-dime" criminals into moving to another area. Secondly, traffic stops by alert officers can get more illegal drugs, illegal weapons and wanted fugitives off the street than any other single method of enforcement. Traffic stops in high crime areas can result in criminal arrests on about one out of four stops. A 1996 study conducted by the International Association of Chief's of Police found that more than 40% (by weight) of all illegal drugs seized in the United States are found during traffic stops. Considering the massive resources allocated to drug enforcement at our borders, airports, customs checkpoints and in local, state and federal investigative operations, 40% is a significant amount. The same study found that the number one police activity resulting in illegal firearms seizures is traffic enforcement.

Once the traffic patterns are rendered manageable by license checks and/or the use of area warrants, aggressive enforcement of moving violations should be started. Careful review of the traffic code will quickly reveal a wide menu of laws that are routinely ignored by gang members and drug dealers. In each state there are code sections that allow or require towing of the violator's vehicle. There are also code sections punishable by stiff fines. The loud-music statute has been a good tool in the Bibb County area. Enforcement of the state loud-music statute has been particularly important as loud music near school has been found to be very disruptive to the learning environment.

Officers should view each stop as a potential felony arrest, which can take a drug dealer or gang member off the street. They should also remember that by finding convicted felons in possession of firearms, they actually help to prevent homicides.

It is important that officers be courteous and professional during stops while also being alert and inquisitive. Each officer may develop his or her own style but they must learn to grasp the key concepts to maximize quality arrests without alienating the general public. In other words, tact is just as important as technical proficiency in these operations.

Officers who master the concepts of effective traffic enforcement capitalize on remaining vigilant for clues that provide justification for a lawful search. Again, officers should be willing to request and act upon lawful consent to search vehicles stopped for traffic violations. These vehicle searches should be quick but efficient. Officers should be alert for weapons and drugs in hidden compartments, under consoles, in sunroof slots, under the dashboard, inside seat springs and in other frequently used hiding places. Officers must remain current on the ever-changing case law on consent searches of motor vehicles. In addition, legislation has been proposed to prevent "profiling" of specific groups by police due to allegations of misuse of consent searches. As with any search and seizure guidelines, officers must be respectful of the rights of citizens and not abuse their authority.

The next tool for cleaning up high crime areas near school is heavy enforcement of laws and ordinances concerning loitering. Some states have specific loitering codes focused on school safety zones. Making cases of this nature can deal with those street criminals who cannot afford a car or who choose not to use a car. Some communities have enacted local ordinances to address this problem when no relevant state law exists.

In some cases, the problem seems beyond solution due to large numbers of individuals loitering in school zones. The old adage "shoot one in front of the others" applies here. Of course, we don't really mean that gunplay will be utilized! What we do mean is we want to take advantage of highly visible arrests. We want word of mouth to clear the streets for us.

A team of officers should be used to approach each individual and/or group loitering near the school. Those individuals that cannot demonstrate adequate legal reason to remain are advised to leave the area and informed that they will be arrested if they are found loitering in the area in the future. Officers should conduct as many pat downs for weapons as legally possible during these stops. This will help to make the operation safer for officers and will help get the word out on the street that the school zone is a risky place to have an illegal weapon. Typically, officers will find that some of the individuals will return the next day.

As officers cannot recognize all violators, they should pick out a few that they are certain were present on the previous sweep. Barring a reasonable explanation that establishes a legitimate need to be present, such individuals should be quickly arrested, cuffed and removed from the area. The remaining hooligans should be told that each individual who is recognized on subsequent days will meet the same fate. A camera may be used for group photos to provide better documentation. These operations will normally reduce several hundred loiterers to a much more manageable group in a very short period of time. At that point, the use of warning citations and arrests by citation may be adequate to keep the area cleared. If your state does not have a state statute of this type, a local ordinance can be enacted to address this problem.

The final common method to clean up a rough area is the use of observation arrests. This can be done with uniformed officers, plainclothes officers, marked patrol cars, unmarked patrol cars, foot patrol, equestrian units or mountain bike patrol officers. A very effective method is to use combinations of the above such as uniformed officers in marked patrol cars working with uniformed officers on mountain bikes. Another very effective technique is to use concealed officers equipped with high-powered binoculars to spot violations in conjunction with uniformed officers detailed for apprehension of suspects.

Officers assigned to make observation arrests should be given specific instructions for the most efficient use of time and resources during peak periods such as just before, during and after school dismissal. Officers should be focused on making as many quality observation arrests as possible during these periods. Observation arrests often pertain to drug offenses, alcohol violations, weapons violations, sale of stolen goods and harassment of students trying to leave school. Officers in this role can make arrests for crimes they witness such as the sale of illegal drugs and for arrests that result from stop and frisk situations that arise from observation of suspicious activities.

The most frequently utilized form of observation arrest will involve the use of officers in vehicles. These officers are freed from all other duties and patrol a specific confined area in an intensive manner. Officers should be alert to any violations that could result in charges being filed. The situation will dictate whether marked or unmarked cars are best. The key is to ensure that offenders in the area are being arrested vigorously. This activity can get the word out on the street that the area should be avoided by criminals.

In combined surveillance and apprehension operations, the key is to make street criminals feel that the police may be present even though they are not visible. Surveillance officers can be concealed in a variety of ways that are limited only by the creativity of the officers involved. They can observe a targeted area from the rooftops of high buildings, from inside vehicles with tinted windows, from inside vacant houses or buildings or from wooded areas using camouflage clothing to conceal the officer.

One important aspect of these operations is that suspects are quickly approached and arrested by the apprehension team after the surveillance officer witnesses the violation. Suspects are initially stunned to be arrested by officers that they know could not have witnessed the violation. Suspects are then informed that they were observed by concealed officers during the commission of their crime. Suspects should never be given specifics as to how and where the officer was concealed during the operation. Not revealing the officer's location will create the desired paranoia among street criminals who now will never know when they are being observed by hidden police personnel.

Another variation of the same technique is to utilize surveillance cameras to spot violations. This method often allows one spotter to observe activity in several areas at the same time. Cameras may be concealed or in plain view depending on the desired effect. Concealed cameras keep street criminals guessing but do not have as much deterrent effect on a day-to-day basis. Visible cameras provide the daily deterrent value but street criminals may be able to move to areas that are not covered by the cameras.

The safe perimeter concept is equally important for athletic events and other after-hour activities. As these functions are often held at night, a secure perimeter can be even more critical for conduct of a safe and peaceful event.

As an example of utilization of the perimeter concept, let's look at how perimeter problems were solved in the author's jurisdiction. From 1989 to 1996 there were two people shot and killed on city streets adjacent to Bibb County Public High Schools. There were an additional six shooting incidents where fatal injuries did not occur near middle and high schools. There were a number of other assaults, some involving weapons. During the same period, officers of the Bibb County Board of Education Campus Police Department interrupted four gang-related shootings that were about to take place. In each case guns were seized and available information indicated that a shooting had been imminent. All of these events occurred off school property while students were present on campus. During the same time frame, on-campus weapons seizures had dropped by more than 70% and overall incidents were down by more than 80%.

During the first four months of the 1996-1997 school year, there were an alarming six shooting incidents on city streets adjacent to county high schools. In one incident, bike-mounted officers from the Campus Police Special Operations Unit rode into a gunfight involving approximately 25 members from two rival gangs and broke up the gunfight, seizing two handguns and arresting three gang members. This incident occurred within 500 feet of a high school just as students were being dismissed for the day.

At the same time, large numbers of gang members began gathering near several high and middle schools each day. Additionally, hundreds of gang members, drug dealers and other street criminals were repeatedly loitering around several schools during peak times each school day. At Central High School it began taking more than one hour each day for students and parents to clear the campus each afternoon compared with 15 to 20 minutes at the other high schools.

These problems were directly related to a dramatic growth in gang activity in the community. The problems were solved through aggressive measures and teamwork involving a variety of agencies and key individuals.

The 1996-1997 school year was also Dr. Gene Buinger's first as the system's new school superintendent. Dr. Buinger approved the immediate expansion of the seriously overworked Campus Police Department by adding 14 contract security officers as an emergency measure. These officers were added at that time to relieve 18 sworn officers by handling all duties not requiring an armed and sworn officer.

Next, there were a series of productive meetings between Dr. Buinger, the author, Macon Mayor Jim Marshall and his newly appointed police chief John Vasquez. Under past administrations, the Macon Police Department exercised primary enforcement in the school safety zones. Even though school police officers had authority to enforce the law off of school property, officers had only exercised this authority in emergency situations or when working directly with city police officers. This limitation of operational scope had been agreed upon by previous civilian administrations and had been in place since the creation of the Campus Police Department in 1972. It was decided that Campus Police Officers would begin to exercise city-wide jurisdiction after being cross-sworn as Macon Police Officers. Campus Police Officers would still operate primarily on campus, but would exercise joint jurisdiction with the Macon Police Department in the 1000-foot school safety zones. Mayor Marshall ordered Chief Vasquez to swear in all officers from the campus police department.

Beginning in February of 1997 a series of dozens of license and insurance checkpoints were conducted on city streets near area high schools. Some were conducted jointly with the Macon Police Department and/or the Bibb County Sheriff's Department, but the majority were conducted solely by the Campus Police Department. Three bike officers from the nearby City of Perry Police Department, detailed to the Campus Police Department for one week of advanced on-the-job training, also assisted with these operations. During these license and insurance checkpoints, hundreds of citations were issued and numerous arrests were made. This resulted in a greatly reduced traffic flow around the high schools. Local television news stations were contacted and asked to report on these operations to inform the public what was being done and why the measures were being taken.

After the traffic in the targeted areas was significantly reduced by the use of the license and insurance checkpoints, an aggressive traffic enforcement program was initiated focusing on moving violations. Traffic stops were made around secondary campuses with officers moving daily to different locations. License and insurance checkpoints were continued on occasion to supplement the moving violation enforcement. Initially, one arrest resulted from every six traffic stops made. Eventually, the arrest ratio dropped to one arrest per 200 stops as street criminals learned to avoid the targeted areas.

In the last 90 days of the school year, campus police officers issued more than 2,000 State Court traffic citations with an average fine amount of $200 per citation. These operations made a dramatic impact on the targeted areas.

Support for a quality school safety strategy must come from the top. Dr. Gene Buinger provided solid support for the Campus Police Department and helped the department expand in size and scope of operation while he served as School Superintendent of Bibb County.

Officers wrote tickets for a wide range of violations, but certain pertinent offenses were targeted. Loud music violations were aggressively targeted due to the disruption of the academic environment that they cause. Another common citation was for violation of the state's window tint code section. This type of enforcement was viewed as important because drug dealers and gang members often use illegal window tint to enable them to handle drugs and firearms without being seen from outside the vehicle. Gang members have been known to use window tint to allow them to ready their weapons before a drive-by shooting, rolling down the windows just prior to opening fire. Several gang members were arrested for illegal possession of handguns as a result of traffic stops for window tint violations. One handgun was recovered that had been used by the suspect to commit a drive-by shooting in a school zone at night, one week prior to his arrest.

During these operations seven handguns were recovered from traffic stops and another 12 felony drug arrests were made. One loud music stop resulted in the recovery of a wanted fugitive with more than $5,000 worth of crack cocaine concealed in his underwear. Hundreds of other arrests were made for driving under the influence, driving on a suspended license and outstanding arrest warrants.

The other techniques that have been described were utilized to clear undesirable elements loitering on foot in the targeted areas. After a number of arrests for loitering in a school safety zone and for observation arrests, the streets became much safer for students who walk to and from school. Group photographs were utilized to record which individuals were being warned not to loiter in the zones so they could be arrested if they returned. The Macon Police Department/Bureau of Alcohol, Tobacco, and Firearms Gang Task Force along with the Bibb County Sheriff's Department Crime Suppression Unit provided valuable assistance in these operations.

During the second half of the school year, there was only one shooting in a school safety zone and that incident had taken place before the area was targeted. Campus clearing time at Central High School was reduced to 15 minutes as cruising in the area dissipated. Despite criticism from the local newspaper, the operations were a dramatic success. The actual and perceived level of safety on and near secondary campuses was greatly enhanced by these operations. During the fall of the 1997/1998 school year, the areas were maintained by periodic use of these methods but arrests and traffic citations declined greatly as the desired effect had been achieved.

This example illustrates that while these operations can take a great deal of effort and coordination, they can be very effective at solving significant problems that at first glance might seem insurmountable. The almost remarkable reclaiming of previously crime-infested neighborhoods close to schools is well worth the commitment of resources required to enhance the safety of students and staff who have to move through these areas. For schools in neighborhoods that have severe crime problems, a number of school districts have constructed "bullet walls" to protect school children. Cleaning up the neighborhood through intensive enforcement efforts may be a less expensive and more meaningful strategy.

For those who do not have these types of problems near their schools, it should be quite apparent how fortunate they are. It should also be apparent that this type of situation should never be allowed to develop. Again, it would have been a very simple matter to be proactive and keep the situation from developing rather than to be forced to react as was the case in this scenario. To further demonstrate the importance of these efforts, after the author left Bibb County to take a state level position, a new school superintendent decided to stop all off-campus enforcement by campus police. This resulted in one shooting on the edge of a campus and another shooting incident near a campus in just one year.

2. Vehicle checkpoints on campus

This method frequently also gets the weapons prior to their entering the school building. License checks like those already described can also be utilized in high school parking lots with good results. As in off-campus operations of this type, general safety is also improved.

Another related method that can be utilized is to conduct random vehicle inspections like those done on military installations. For this type of operation to be legal, prior notification through signs, posters and other means should be given before such tactics are used. For this effort a search area and team are set up and randomly selected cars are stopped and checked for weapons. Documenting that the selection process is truly random is a critical legal point.

Although the author has not personally used this technique, other school systems have implemented it with positive results. The concept here is that randomly checking a few cars periodically will deter most students from having a weapon in their car. This tactic would not be suggested unless a chronic problem with weapons in vehicles exists.

3. Plain view vehicle checks

Plain view vehicle checks are an easy and effective means to reduce the number of weapons brought to school in student vehicles. This method is also less intrusive than random vehicle inspections. Knives, guns, impact weapons, drugs, alcohol and other contraband can often be spotted in student vehicles by officers walking through the parking lot looking through car windows. Typically, only a part of the forbidden item will be visible. Once a banned item is spotted, however, the car can be legally entered, the item seized and the entire

Plain-view checks of student vehicles should be conducted at any school where students are allowed to park on campus.

car searched for related items. Normally, the driver of the car is asked to open it. Most students will readily comply with an officer's request to open a vehicle under these conditions.

It is important that all cars found to obtain a single illegal item be searched thoroughly. Often, additional items will be found. It is also important to attempt to take a written statement immediately from any student found to have a weapon in his or her car. Often, a parent or other third party will claim to have left a weapon in the car without the student's knowledge. Although this will be true in some cases, taking a statement quickly may eliminate the very real possibility that a false story will be concocted later to cover for the student.

4. Canine checks of vehicles

Canine checks of vehicles can be another effective tool to keep weapons out of the parking lot. Dogs (or swine) that have been trained to detect the odor of firearms and ammunition can be used to check vehicles for hidden firearms. Any dog trained to detect explosives can be used in the same fashion and are often more readily available. In this type of program, randomly selected vehicles are checked by walking the dog around the exterior of the vehicle. While even the best dogs cannot always locate a gun hidden inside of a car, they very frequently can detect guns and even ammunition. High frequency and low volume are again a very effective means to deter violators. Ten to 20 cars in a row are selected to be checked by the dog. If the dog alerts on a particular car, it is searched for firearms and ammunition. In areas of the country where hunting and shooting are popular, some alerts will result in ammunition that has been inadvertently left in the vehicle after target shooting or hunting. While these finds do not normally constitute a criminal violation, they are very important in deterring gun violations. When the involved student tells friends that the dog found ammunition in this manner, students become aware that if a gun is left in a car, it may be found by the dog.

A number of school law enforcement programs have followed Bibb County's lead by utilizing gun detection dogs. Officer David Morris from the Department's Special Operations Unit is about to conduct a check of student vehicles with "Josh," the first gun detection dog put into service in the state of Georgia.

By checking a limited number of cars, the dog can be used to check cars at numerous campuses without becoming fatigued. This method is more effective than checking all cars at just one location. Again, the idea is to deter would-be violators rather than to catch violators after a gun comes to the campus. The use of dogs in this manner is a high profile and effective way to prevent gun violations in the student parking lot.

Passive alert dogs are best suited for this type of work. By sitting or lying down to indicate the presence of firearms or ammunition, the dog will not scratch student vehicles like a dog that indicates by pawing. As with drug dogs (which can also be used in this manner), the alert is in response to the presence of the targeted odor. It is important to understand that detection dogs alert to the presence of contraband odor which means that in some cases, a dog will alert to residual odor even though the contraband item is no longer present. In cases where a dog alerts by clawing, there can be minor damage to automotive paint. If the dog has alerted on residual odor and no contraband is found, there may be a need for the individual to be compensated for damage to the car. Passive alert dogs eliminate this potential problem.

It is important that students understand what type of dogs will be used to check student vehicles to make maximum use of the deterrent value of the program. Most people are not aware that dogs can be used to locate guns and they will presume that the dogs are only being used to locate drugs. Again, we want to deter the violator from bringing a gun to the campus rather than to catch violators after the fact.

5. Ground/area searches

These searches are important as student violators will sometimes try to get around other countermeasures by hiding weapons under a bush or other stationary object, or by burying the weapon outside the school. Detection dogs can effectively cover large areas in a relatively short period of time and are very useful for this task. It is important to understand that the dogs are subject to fatigue and are best utilized for short periods of time at various locations. Travel time from one site to another can help to provide recovery time for the dog.

One limitation on the use of dogs in this manner is that they can only find guns. While many police dogs are trained to locate recently handled items of any types, these dogs may be of limited use to deal with this type of threat. Dogs trained in "article-find" can locate guns, knives, drugs and other items hidden by students, if used quickly after their arrival at school. One problem that will be encountered with this type of detection dog is that the dog will alert on any item that has been dropped on the ground such as scrap paper and gum wrappers.

Another method to screen for hidden weapons is to use the same type of metal detector that is used by hobbyists to find buried coins. These detectors can find guns, knives, knuckles and other metallic weapons that have been hidden or buried outside of a school.

Officer Marty Scott uses a Garrett ground search unit to check for hidden weapons at the Bibb County Alternative School. Ground search units can provide a valuable visible deterrent to potential violators.

The best solution is to alternate using a combination of methods. A good way to supplement these screening strategies is to conduct periodic surveillance of students as they arrive at school. Officers using good quality binoculars can see students hiding weapons or other contraband and apprehended by other officers in the area.

Announcing that the preceding methods will be utilized can be an effective deterrent to students who might otherwise attempt to bypass screening measures by hiding weapons outside the school.

6. Entry-point weapons screening

When done properly, this type of screening can be one of the most effective screening measures, but it is by far the most expensive and labor-intensive form of weapons screening. Entry-point screening is what most people think of when the topic of reducing weapons violations in schools arises. The public tends to make this equation since entry-point screening has worked so well to reduce hijackings on commercial airlines. It is important to understand several key points when making this comparison:

- Because commercial flights are a for-profit enterprise, the enormous expense of effective perimeter security and entry-point screening can be charged to the consumer.
- Airport screening is primarily concerned with preventing guns from being carried onto aircraft. In a school setting, knives are a much more common threat.
- The delays caused by screening at airports do not cause the types of safety hazards often experienced in the school setting. One major concern is extended delays caused by entry-point screening that can make students

wait outside a school. This can create hazards for the students, particularly in the dark in early morning hours. Another consideration is the vulnerability of large groups of students waiting in a common area to be screened. Such a situation could create a prime target area for a planned shooting incident.

- Airports are designed or have been remodeled for perimeter security. Schools on the other hand with numerous access points are often very difficult to secure .
- When an individual is caught attempting to smuggle a gun through the airport screening process, he or she faces a certain prison sentence. This powerful deterrent is not always present in the school setting.

Even given all of this, entry-point screening can be a practical and effective measure to reduce weapons violations in school systems that are willing and able to afford the significant expense. A quality entry-point screening program takes a great deal of commitment requiring top quality equipment, top quality personnel, careful planning and adequate time to screen students properly. Note: for a detailed explanation of the technical aspects of the setting up and operation of walk through metal detection equipment see *School Security Screening With Metal Detectors-a Planning Handbook*, which may be requested from Garrett Metal Detectors at no cost via facsimile at (972) 494-1881.

Some school entry-point screening programs fail to be effective because there is not an adequate administrative commitment to making the program work. Often, entry-point screening programs become little more than a facade to placate the public following a serious weapons incident. When weapons continue to show up inside schools that use entry-point screening, this lack of commitment to quality program results in difficult questions arising.

Entry-point metal detection is often necessary to keep weapons out of high risk facilities such as alternative schools.

High quality equipment is a must for effective screening. Because equipment costs are actually quite small in comparison with personnel costs, it can be an expensive mistake to fail to buy quality equipment. Certainly, it is wasteful and expensive to replace metal detection equipment after finding out that it is not adequate. It can be even more expensive in the long term when personnel costs must be increased to make up for equipment shortcomings.

Quality personnel are also critical. Mediocre people will yield mediocre results. The author once passed through a weapons screening checkpoint at a college graduation while carrying a fully loaded police service pistol and an extra-high-capacity magazine. Although the detector properly alerted, the police officer working the checkpoint assumed that a small umbrella had set the unit off. The author had intended to show his credentials, but was never questioned by the officer who could have just as easily allowed a dangerous individual to pass with a large handgun and 40 rounds of ammunition.

Without a carefully thought out screening program, it may be a very simple matter for students to notice a weak spot that can be exploited. Methodical planning can reduce potential weak spots that might otherwise render entry-point screening almost useless.

Adequate time for quality screening is also important. In many situations, personnel are urged to speed up the screening process because of a backlog of students. When people are screened too quickly, the entire process can be seriously compromised. An important factor in avoiding delays without sacrificing quality is the number of detection stations and the number of personnel assigned to screening duties. Again, proper planning can keep the students moving through the screening process while maintaining quality in the process.

The programming of the walk-through metal detector will determine the amount and type of metal that it takes to trigger the alarm on the detector. It is important to understand that different weapons are made of different quantities and types of metals. Decisions must be made concerning the most appropriate setting for the detector. A more sensitive setting will result in a greater number of alarms. Of course, a wide range of weapons will be detected on these more sensitive settings. A balance between the available time for screening and the type of weapons to be screened for must be achieved. One highly successful strategy is to adjust the sensitivity setting from time to time to keep would-be violators "on their toes". This tactic is designed to create doubt in the minds of those who might try to sneak a weapon through the checkpoint. Newer, multiple zone detectors can help to speed up screening by pinpointing where on the body the metal is located. These units are also less intrusive as the whole body does not have to be checked with a hand held metal detector to locate the metallic item(s) that trigger an alarm.

An entry-point screening station should be designed to guide the flow of people to be screened in an orderly fashion. Simple physical barriers should be used to keep students moving to the walk-through detection units while preventing any of them from walking around the units. Personnel working the station should have specific duties that do not overlap to result in gaps during peak activity levels. Flow through the walk-through units should be steady but at a pace that keeps each individual separate as they enter the detector.

A separate screening area adjacent to the walk-through unit must be established for checking those individuals who trigger an alarm when they pass through the unit. Each must be checked carefully with a hand-held metal detector to determine precisely what caused the alarm. This separate function is needed to prevent the entire process from coming to a halt each time the walk-through unit indicates an alarm. In situations where only a small number of individuals are being screened (such as an alternative school) the two stations can be combined. When this is done, it is critical that the flow of students stops completely whenever the officer's attention is diverted by hand-held screening.

An important consideration is the screening of books, book bags, purses and any other hand-carry items. Some school systems utilize x-ray units to screen such items. This method is less intrusive, but it is also extremely expensive and can miss certain types of weapons. While x-ray equipment is a very effective means to screen for handguns and explosive devices, when personnel remain alert, it is not always as effective as hand searching the items. Another advantage of manual searches is the increased likelihood that properly trained searchers will find aerosol dispersal devices that have been improvised to become weapons of mass destruction. The main advantage of the use of x-ray equipment is much easier to justify in court than manual inspection.

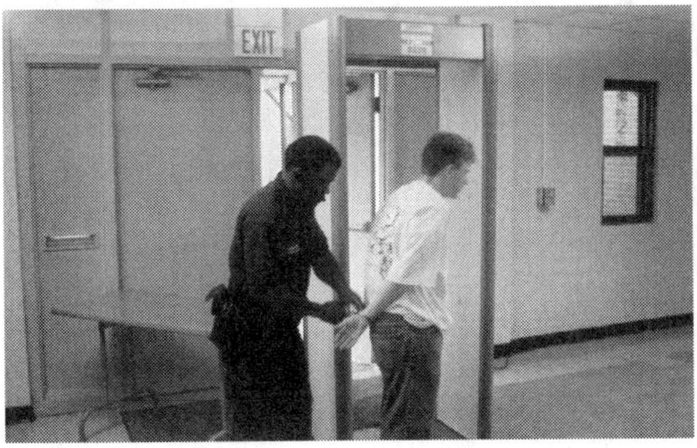

If a student is caught trying to bypass the checkpoint with a firearm, they should be immediately secured. The type of violator who tried to bypass a checkpoint with a weapon is often a high risk offender who may assault the officer when caught.

Manual inspection of hand-carry items can be done while students are checked by the walk-through detection station. Personnel assigned to this task must work quickly and efficiently to screen items adequately while maintaining the flow of students through the detector. Policies that restrict non-essential items eliminate many items that would normally have to be inspected. A table placed near the walk-through unit can serve as a platform on which hand-carry items can be inspected. Personnel assigned to these duties must remain particularly alert for dangers posed by hypodermic needles and razor blades.

An important consideration is the positioning of an armed and sworn officer

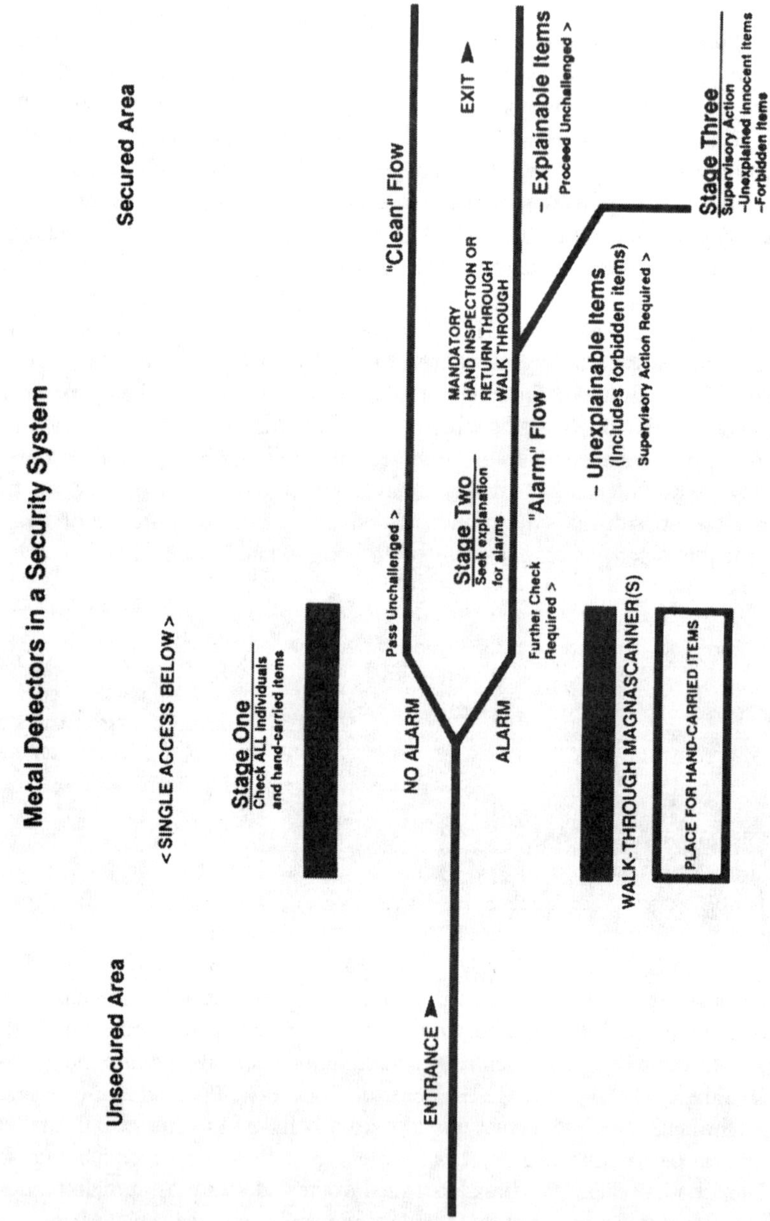

Metal Detectors in a Security System
Unsecured Area
Secured Area
< SINGLE ACCESS BELOW >
Stage One
Check ALL individuals and hand-carried items
ENTRANCE >
NO ALARM
ALARM
Pass Unchallenged >
"Clean" Flow
EXIT >
Stage Two
Seek explanation for alarms
MANDATORY HAND INSPECTION OR RETURN THROUGH WALK THROUGH
Further Check Required >
"Alarm" Flow
- Explainable Items
Proceed Unchallenged >
- Unexplainable Items
(Includes forbidden items)
Supervisory Action Required >
Stage Three
Supervisory Action
-Unexplained innocent items
-Forbidden items
WALK-THROUGH MAGNASCANNER(S)
PLACE FOR HAND-CARRIED ITEMS

in close proximity to the screening station. This is important in case a student is detected trying to smuggle a weapon through the checkpoint. If such an officer is not present, a weapons assault or even a hostage situation could result when the violator panics upon the detection of his or her weapon. And, the presence of such an officer will serve as a deterrent to these extreme situations, adding greatly to the powerful psychological impact of the screening station itself.

The actual screening process can be performed by security personnel or even properly trained school staff members to help reduce costs. It is critical for the safety of all involved, however, that a uniformed authority figure be present at each screening station, especially during the screening process. In most cases, the combination of security personnel and one police officer is the most practical method.

Careful thought should be given to these tactics that could be used by students to try to bypass entry-point screening:

- Placing a weapon on a window ledge outside a classroom. After being screened, the student opens the window and retrieves the weapon.
- Bringing a weapon into the building after hours when screening is not being utilized. The student then hides the weapon in the building to be retrieved when needed.
- Utilizing a technique called "misdirection" to get a weapon through the screening checkpoint. In this tactic, the violator hopes that the operator will accept an innocent item as the reason for the alert and not inspect more closely to locate the forbidden item. Just as the operator assumed that the author's umbrella triggered the alarm and did not detect the author's Glock pistol, a violator may utilize a large belt buckle or an item of jewelry to serve as the apparent reason for the alarm. Personnel training by Garrett emphasizes that all alarms must be properly investigated with the actual cause of the alarm determined.
- Disguised weapons may be utilized by a violator in the hopes that the operator will be fooled by a its disguise as an innocent item. A student at the Bibb County Alternative School was arrested when the officer carefully examined a crucifix on a key chain that triggered an alarm. Inside the crucifix was found a small knife blade. The offender obviously thought that he could "beat the system" but learned the hard way that the "system" and the officer were prepared for this type of threat.
- A person who has not yet been screened passing a weapon to a person who has already cleared the screening station. This tactic may succeed if the screening station is poorly laid out, and/or is not adequately staffed.

Most of the potential weaknesses in entry-point screening will involve perimeter control or operator error. Physical control of all potential entrances and openings to the building and/or grounds must be maintained. This can be done with personnel and/or video camera surveillance systems and exit alarms. The use of video systems can be more expensive initially, but is often much more cost effective in the long run.

Entry-point screening for weapons will be most effective when supplemented with the other measures that have been discussed in this chapter. A particularly effective strategy is to conduct efficient entry-point screening along with occasional random screening. When the two methods are utilized together, the optimum deterrent value will be achieved. Effective entry-point screening is the best countermeasure to the "plotter" who is determined to commit a serious weapons assault. This is another reason that at least one armed police officer should be on duty at each entry-point screening station at all times. If an armed officer is not present, the offender can simply plan to shoot his way through the screening station. And of course, the other countermeasures discussed in this book must be present to keep the offender from simply shifting the assault location to the parking area or other public area outside of the screening station. Of course, without entry-point screening this individual has a much greater chance for success because the odds are that random screening will not effect this particular student on a given date.

If done properly, entry-point screening can be the most effective measure to prevent weapons from entering a building. As long as potential weak spots are addressed through adequate countermeasures, entry-point screening will be what the general public expects it to be. Of course, local conditions will dictate whether entry-point detection is the best use of available funding to provide the desired level of protection. In some cases, random screening programs are a more appropriate method to reduce weapons violations.

7. Random Screening

For many school systems, surprise screenings of randomly selected groups of students provide the most viable option to deter students from carrying weapons on their person. Random screening eliminates the problems of perimeter security: significant additions of staff and equipment, serious delays of students entering school and the fiscal burden continually required by an effective entry-point screening program. A proper random screening is less intrusive as students are only checked on occasion. Of course, random screening does not provide many benefits that can be provided only by a quality entry-point screening program. Again, the most effective approach is the use of both countermeasures together.

The concept of surprise random weapons screening is designed to create an academic environment where the day to day routine is not disrupted unnecessarily, while at the same time deterring most students from bringing weapons to school. This is another type of preventive strategy where high frequency/low volume is a key to success. For this method to be legal and effective it must be reasonable for the threat level and must be truly and demonstrably random. At least one school search-and-seizure legal expert feels that the use of random metal detection is now considered to be reasonable in any community because of the number of school shootings that have occurred in both rural and suburban settings.

This countermeasure involves a properly trained search team with at least one armed and sworn officer present at all times. This is critical to counter the potential threat of a major weapons incident if a student becomes desperate when caught with a weapon. The random search team conducts a search of groups of students chosen by a lottery selection process. The students are

When conducted in an sensitive and caring manner and when the public is provided appropriate background information regarding the dangers of weapons in schools, random screening is not only effective, it is often popular with students and parents.

normally selected by computerized or manual drawing of classroom or school bus numbers. These searches are performed at various times so students can never guess when they might be subject to inspection. Checking busloads of students as they depart a bus in the morning or get on a bus in the afternoon helps to deter these students from hiding weapons upon their arrival at school each day.

Random screening of students as they arrive at school or prepare to depart from school on buses can also be a very powerful deterrent. School bus drivers are naturally strong supporters of this technique.

Random searches can be conducted using hand-held metal detectors or even walk-through units. Each school poses unique requirements that will help to determine which type of equipment is best for the situation. Many students and parents will prefer that walk- through units be employed since they may seem less intrusive than hand-held instruments. Most search teams will prefer the use of the hand-held units because of their portability and precision in locating small metal items such as razor blades.

For random searches using walk-through metal detectors, the unit is set up in a central location in the building and the selected groups of students are brought to the unit to be screened. In this method, the team enters the selected classroom and informs the students that they have been randomly selected for weapons screening. Students should be instructed to keep their hands out of their pockets until they have been screened. The team members must be very alert for violators who try to dispose of contraband items.

Most contraband is recovered because violators will often telegraph the presence of the item by physical actions. Just as a speeding motorist tends to hit the brakes when they see a police car parked on the side of a highway, the weapons violator will often make a sudden movement of one hand to the concealed item's location. When this occurs, the student should be asked if he or she has any contraband on their person and if they do not identify the item, they should be taken into the hallway for further inquiry immediately. It is also a good idea for room numbers to be selected prior to the team's arrival on campus and for the team to move quickly from room to room. If students feel that they will have an opportunity to dump weapons before or during random searches, much of the deterrent effect will be lost.

The students are next instructed to move into the hallway in an orderly manner to maintain adequate observation by the team. They are then guided to the detection station and screened in similar fashion to entry-point screening. Care must be taken so that students who have not yet been screened cannot pass contraband items to someone who has been screened. Team members must also remain alert to the possibility that a student who is carrying a weapon might try to dispose of the weapon by breaking and running. Once the students are screened, all book bags, pocketbooks and other hand carry items which were left in the classroom are carefully inspected. If enough team members are available, such items may be checked while the students are being screened.

When using hand-held metal detectors instead of walk-through units, students can be screened in the hallway next to the classroom. This reduces the logistical problems of moving students to stationary walk-through units. It is important that officers are trained in proper use of the hand-held units. Officers must be consistent in screening all physical body areas of students being screened.

With either type of random screening, no students are exempt from the

Students should be scanned by personnel of the same sex to avoid making them feel uncomfortable about being screened. It should also be stressed that officers should conduct themselves in a friendly and professional manner, showing respect for students at all times.

screening process. Any student who does not wish to be scanned with a metal detector should be given the option to be searched by hand. Any one who requests this should then be searched very carefully since they may be trying to avoid being caught with a weapon.

At the same time, it is important to understand that a very few people are truly afraid of being scanned by a metal detector. Some heart patients may believe that a metal detector may shut down a pacemaker or other device, but studies fail to support this possibility. Even so, trying to convince a concerned student that screening is safe may be fruitless. It is probably simpler to search the student by hand. This might also help deter other students from falsely declaring that they have a medical problem just to avoid being checked for weapons.

After the search is completed, students should be briefly reminded that the search was conducted as a result of a random selection process and that searches are done for the safety of all students. Students should be reminded that a single student with a gun on campus places all students in great danger. It is important that students buy into the random search strategy and taking a minute to address each class is time well spent.

A good example of the need for officers to be alert during such operations occurred when a Bibb County School District Police Officer noticed a satanic symbol on a student's key chain during a random weapons check. When the officer and school officials continued to inquire about this symbol, they developed cause to search the student's locker and found satanic writings and, more importantly, a detailed suicide pact between the student and another student at the school. Both were placed in a mental health facility for treatment after their parents and psychiatrists assessed the situation and determined that the students had truly been determined to commit suicide. Because two high school students had committed suicide with a handgun in a Georgia high school bathroom several months earlier, there was concern that this suicide could have taken place at school. Regardless of where the suicide would have occurred, this random search averted two teenage deaths.

Even though drugs will often be discovered in such searches, it is a critical legal point that all involved understand that the searches are actually focused on the location of weapons, not drugs. Courts in most areas of the country have held that random weapons searches are not an "unreasonable" type of intrusion given the significant danger that a student with a weapon poses to other students. The courts have also routinely ruled that random searches for drugs alone will not be upheld because the danger to other students is not great enough to justify the intrusiveness of these searches.

Every member of the team should be reminded that the courts have generally been supportive of school officials in allowing random weapons searches. Officers must understand that the problem of weapons on campus is too serious to jeopardize the valuable tool of random weapons searches by trying to "push the limits" looking for drugs. Drugs will be found on occasion during lawful random weapons searches, but these same drugs would not have been found if officers had abused the law leading to a court order that the school system discontinue all searches. In short, weapons searches will help to reduce drug violations, but the drug problem will require other measures.

An experienced team can conduct a classroom search of 20 to 30 students in 10 or 12 minutes, normally allowing a team to arrive, search two or three classrooms and leave the campus in one class period. Of course, if weapons are found, the team will be slowed while an offender is processed. If an officer is regularly assigned to the school, it may help to have this officer accompany the team and handle any arrests. This will serve to keep the team moving when weapons are found. This is particularly important in larger school systems where the team plans to conduct a series of random searches at various campuses.

While it is normally desirable that the team only be on the campus for a short time and then quickly depart, it is also important that the process be unpredictable. One way to accomplish this is occasionally to conduct searches several times on one campus during different periods during the same day. Searches may also be conducted two days in a row at the same campus. These tactics are designed to keep students from deciding to carry a weapon on days immediately following a search.

In communities where gang activity is prevalent a useful concept is to make it a practice to increase the use of random weapons screening at a school when there is a great deal of tension between rival gang members, whether at the school or in the neighborhood. As students discover this common practice, the chance is reduced that a student will bring a weapon at a time when such violations are more likely to take place. This is particularly important following a gang-related shooting that involve students, even when the incident has occurred off campus.

Random classroom searches will typically cause a dramatic reduction in the number of weapons being carried to school. In some cases when the searches are first used, weapons violations may seem to be increasing because weapons are located that would otherwise have not been found. In most cases, the actual number of confiscated weapons will drop once the searches have been in place. In every case, the actual number of weapons on campus will drop.

Random searches are a cost effective means to reduce weapons violations, and this method will not result in students spending as much time being screened as required by entry-point inspection. Again, the combination of entry-point screening and periodic random weapons screening will provide the ultimate level of safety.

8. Random locker inspections

Lockers are by far the easiest hiding place to deal with in a school setting. Random inspections, sometimes using detection canines, can rid these popular hiding places of contraband. In school systems where lockers are not inspected, a high percentage of weapons and other contraband will be found there. Yet in schools where adequate inspections are conducted, it is a rare event to find contraband in a student locker.

It is important that the school make it very clear that all lockers are property of the school and are subject to search at any time without cause, reducing any expectation of privacy. It is also important that all locks be property of the school with no other locks allowed. Students should be advised that any other lock will be cut off and the locker carefully inspected. Equally important is the principle of keeping locked those lockers that are not in use. This prohibits a student from hiding weapons, explosives or other contraband without accountability in an empty locker.

Some school systems have solved the problem of contraband in lockers by eliminating lockers altogether. Doing away with lockers creates other safety hazards and in some cases results in enormous expenses. If the school system is willing to commit resources for locker inspections, however, denying students the use of lockers is not necessary.

The first type of locker inspection is the random locker search. Officers manually search a group of lockers that have been randomly selected. This is another case where high frequency, low volume is the key. The search of even five to twenty lockers once each week normally provides an effective deterrent to a potential violator. It is therefore important that officers be seen when checking lockers which ensures that students are continually reminded of the danger in keeping a weapon in their locker.

It is important that each locker selected searched carefully. For instance, books should be checked for hidden compartments. When the inspection is completed, all items should be returned to the locker and the locker securely locked. Careful legal research should be conducted to ensure that searches are conducted in compliance with recent court rulings.

It is not unusual for locker inspections to reduce other types of violations. Locker checks can be particularly effective in helping to reduce alcohol violations, theft violations, vandalism (spray paint cans are often concealed in lockers) and to some extent, drug violations. While again, the focus of the searches must be on weapons, a reduction in these other types of violations can be an added bonus.

It is also a good idea to supplement these inspections with periodic use of gun detection and drug detection dogs to inspect lockers. Dogs can quickly check a large number of lockers and are another high profile means of reminding would-be violators that lockers are off limits for contraband. The combined use of dogs and manual locker inspections can effectively eliminate the use of lockers as a hiding place for weapons.

9. Inspections of public areas

Public area inspections inside the school are another important preventive measure. Students will sometimes hide contraband under a vending machine, inside a sink, in a bathroom ceiling or in some other similar location. Here again, the use of detection canines is an effective countermeasure to deal with drugs and firearms violators. Because odor normally drifts downward, it is not unusual for a dog to be capable of detecting contraband at ceiling level.

Another effective method is random inspections of those areas that require using ladders and flashlights. Again, it is important that students be made aware that these inspections will be conducted. The goal is always to deter the violator, rather than simply to find contraband. Students should never be told what specific areas will be checked, rather they should be advised that all such areas are subject to frequent random inspection.

10. Reasonable suspicion searches

Another important screening measure is the search of specific individuals based on reasonable suspicion of those individuals. It is critical that students know that school police can and will search individuals based on tips. If students "buy into" the schools weapons reduction strategy, they will normally be willing to come forward with information regarding students who have not been detected by other measures.

Careless handling of tips concerning students who may have a weapon can not only result in the student "dumping" the weapon, but may also result in injury or even death to innocent people. It is critical that such tips be taken seriously and handled with care. There have been far too many cases of a serious weapons assault taking place because this type of situation was mishandled. In one such instance in Florida, one school administrator was shot and killed, another was shot and wounded, and a student was shot by responding police officers when administrators attempted to search a student for a gun without police assistance. In another case, a mass school shooting occurred after school officials missed a handgun when searching a student without police assistance.

Tips must be evaluated to determine if the legal basis for a search exists. School administrators and police officers should understand that many tips are not completely accurate. In many cases, a tip is received that a student has drugs or even a less serious type of item such as a pager or pack of cigarettes when in reality the student has a gun. Often, a tip regarding a student with a knife reveals a student with a gun or a tip that a student has a gun results in a seizure of a knife or drugs. It is therefore important that every tip be regarded as a possible indication of a student with a loaded gun. First and foremost, therefore, it is strongly recommended that all viable tips be handled by an armed police officer.

An effective strategy is for the school police officer to go to the student's location with a school administrator or other civilian staff member. If the student is in a class, the officer(s) should remain out of sight in the hallway. The staff member should ask the student to step into the hallway in a very casual manner. The staff member should closely watch the student's hands to ensure that the student does not "dump" or pass the weapon to another student. Once the student is in the hallway, he or she should be asked if they have any item that is prohibited. If the student states that they have no contraband, he or she should be told that this must be verified for the safety of all students at the school. If the tip concerns contraband other than a weapon, the student should be asked to empty all of their pockets and then to turn all pockets inside out. Officers should be alert to attempts to hide drugs in the palm of the hand or in contents of the pockets. Officers must be ready to react in the event that the student attempts to flee or to produce a weapon. The search for items other than weapons should be handled in this manner to satisfy legal requirements involving the use of the "least intrusive means" that is practical. After the student has emptied all pockets and turned all pockets inside out, the student should be patted down carefully for weapons.

In tips relating to a weapons violation, the officer should retrieve any weapons that the student admits being present. In cases where the student denies having a weapon, the officer should immediately search the student by hand without asking that pockets be emptied. In cases involving a weapons tip, most courts allow officers to move directly to the hand search due to the inherent danger of allowing the student to reach inside their pockets. It is not a bad idea to check the student a second time using a hand-held metal detector just to be certain that no items were missed during the pat down.

Once the student has been checked, the student's book bag, books, purse or other possessions should be inspected if justification exists. It is also recommended that the student's locker and vehicle be inspected as well. It is important to do so even after the suspected item is recovered because violators will frequently have additional contraband in one of the other locations.

The civil courts will be very unfriendly and unforgiving to school administrators who fail to take information regarding weapons seriously and fail to search when a weapons assault occurs as a result of this failure to act. Also, administrators who mishandle this type of search greatly increase the chances of injury or death to themselves and innocent bystanders.

Aggressively pursuing weapons violators through professional, prompt and thorough response to tips regarding weapons violations is a vital part of an effective screening program. Making students aware that the school will respond appropriately to information that is received can be yet another effective deterrent to the potential violator.

The measures that have been outlined in this chapter, when utilized together along with state-of-the-art metal detection equipment and quality personnel will have a dramatic positive impact on the number of weapons violations in a school system. Although each system has a different threat level that will determine which measures are called for, most schools have a need for at least some of them. Many school systems have a true need for all.

Chapter 6

CAMPUS PATROL METHODS

The modern school campus lends itself to a variety of traditional and not so traditional patrol techniques. And again, each school is unique and may have specific needs not present in other schools, even in the same community or region. It is useful to tailor patrol tactics to solve those specific problems that are present or to prevent difficulties that are likely to develop in the future.

Foot patrol is one of the most effective forms of preventive police patrol and is by far the most common form in use in American schools. This traditional form of patrol has been making a dramatic comeback in many police settings across the nation after its use had become rare. While it may seem the most expensive, at least at first glance, many police agencies are now finding that the cost savings that led to the deletion of foot patrol assignments have also led to a decrease in prevention of crime and to an increase in the need to respond to crimes. Even though one officer in a patrol car can cover a beat that would require four or five foot patrol officers, the quality of coverage is greatly sacrificed.

Foot patrol assignments are effective in a school setting because they allow the officer to develop a much greater degree of rapport with students and staff. Officers on foot patrol duty will also be in a position to see and hear things that will be missed by an officer in a car. Of even more importance in the school setting is the increased ability of the officer to bond with the citizens they serve through the close and personal contact that foot patrol provides.

There are certainly some disadvantages to the use of foot patrol. Inclement weather can be a problem if there are outdoor areas that require coverage. Officers on foot patrol are also limited in the amount and types of equipment that they can have on hand for emergencies, and officers on foot are often unable to stop vehicles. Even with these limitations, foot patrol is often one of the most effective forms of patrol on a school campus.

Officers assigned to foot patrol duties should utilize many of the same techniques that have proven effective in other environments where foot patrol has been utilized. Randomness in officer movement, double-back patrol and patrol in key locations during periods of peak activity will maximize the deterrent value that makes foot patrol so effective. Officers should be particularly aware of known problem areas that can be best addressed by an officer on foot. Continual feedback from school administrators, staff and students should be solicited to ensure that the needs of the school community are being addressed.

Stationary surveillance is another effective means of preventing crime on campus. While this is often an integral function of the foot patrol officer, it is a more specific form of patrol that often does not come naturally to police officers. Surveillance may involve watching a given location from a spot where the officer is clearly visible, from a concealed position such as in an office overlooking a parking lot or from a control room where the officer monitors surveillance cameras during critical times. Some schools have elaborate camera systems that allow one officer to monitor many key areas at one time. Officers must be mindful that camera coverage will normally have blind spots even when numerous cameras are in place. Even with the amazing advances in closed circuit camera technology, it is important not to be so totally reliant on camera coverage that these spots are left unprotected.

Another type of foot patrol that can be very effective on large crowded campuses is the occasional use of plainclothes officers dressed like students. This can be a particularly effective tactic during class changes, at the start or end of the school day and during special events. While students will sometimes recognize officers, it is amazing how often an officer can slip up to a student, group of students or non-student violators in a crowded area. This is yet another "force multiplier" that helps to make violators wonder when the police may suddenly appear.

Bike patrol is another example of innovative police agencies returning to a method that was once heavily utilized. Of course, improvements in mountain bike design, new equipment and new tactics have made this a very effective tool to combat crime in the school setting. Most campuses are ideal places for the use of mountain bikes. These bikes are fast, quiet and allow an officer to cover a larger area more effectively than traditional foot patrol, while still providing a level of quality in coverage that cannot be attained with a patrol car.

A turn-of-the-century mode of policing has been resurrected. As demonstrated by these Washoe County, Nevada School District Police Officers, modern bikes and practical uniforms help them to be prepared to "Assist and Protect Nevada's Future."

Police bikes can be effective in most areas on and around a campus and can be utilized both inside and outside buildings. Students are normally receptive to officers on bikes and will often approach such an officer to ask questions about the bike and its equipment, another reason that this type of patrol is so well suited for use in and around schools. The increased contact between officers and students who might not normally approach an officer is of significant value.

Bike patrol is also suited to after-school activities such as athletic events where bike-mounted officers can reduce problems such as car break ins and the sale of narcotics. The Seattle Police Department found that a bike-mounted officer has a rate of arrest that is five times higher than that of an officer on foot or in a patrol car. Bike patrol is definitely one of the most effective means to deploy a police officer on and around a school campus.

The use of marked and unmarked patrol vehicles is generally not as effective on a school campus as foot and bike patrol, but this type of patrol can provide important coverage. While officers on foot and on bikes can make traffic cases, they simply cannot handle traffic enforcement duties as effectively as an officer in a patrol car. Patrol vehicles also allow an officer to search a large area better than any other means. And, of course, only with a patrol vehicle can an officer handle details like transporting a prisoner, going to court and quickly moving from one site to another to give assistance.

While it does not provide the close contact that is possible with some other types of patrol methods, the properly equipped patrol car enables officers to perform critical functions such as prisoner transport that cannot be done without a vehicle.

It is often a wise use of resources to equip an officer with both a bike and a car. This allows the officer to patrol on foot and on a bike while having a patrol car when needed. Another practical means to attain good coverage is to assign different officers to foot, bike and vehicle patrol in the same area. These officers can then work as a team to provide different types of coverage that can keep potential violators away from a campus.

Another method is to utilize specialty vehicles such as all terrain vehicles (ATV'S) or golf cart-type vehicles which can travel to places where cars cannot. In addition, they can be purchased and maintained at less expense. These vehicles can be particularly useful on large campuses and those with large wooded areas.

Equestrian-mounted patrol is not commonly utilized on school campuses but in certain situations trusty steeds can be worth their weight in gold. Here again, students and staff members alike tend to bond very readily with an officer and the horse. While most school systems do not have enough use for full time equestrian officers, larger systems may find them quite worthwhile. In other systems, members of the local police department's equestrian unit can be detailed to patrol school campuses from time to time.

A variety of patrol methods from the standard to the specialized can meet the specific needs of different campus environments. A thoughtful review of the particular situation can increase chances that the best methods are employed.

Chapter 7
INVESTIGATIONS

Investigation of school-related crimes will normally involve several basic categories. These typically involve crimes committed by students on campus, crimes committed by non-students on campus, crimes committed in school safety zones and occasionally allegations of criminal conduct involving school staff.

This chapter is intended as a general guide to some of the more common and more critical investigative issues of concern to the school police officer. The same basic tried and true investigative methods that work effectively in other environments will normally be just as adequate in the school setting and should be utilized as needed.

The most common type of investigation will usually involve crimes committed on campus, even though most of these incidents are not of a serious nature (felony class violations). Such investigations are often aided by the enhanced relationship that occurs when specific officers are assigned to a particular school. This results in a greater likelihood that students will cooperate with the police than is the norm in a given community. For this reason, it is important that the officer(s) assigned to the school be actively involved with most investigations.

It is common practice in the field of law enforcement to have an investigator take over certain types of investigations and exclude patrol officers. In the school setting, however, it can be more productive to assign school-based police to assist investigators conducting an investigation of a campus incident.

One of the more common types of investigations involving student perpetrators is the assault, normally committed against another student. If an effective school resource officer program is in place at the school, the vast majority of these cases can be effectively cleared. Clearance rates for even misdemeanor assaults will often run into the high 90% range when they might typically run in the 40% to 70% range in the community. Felony assault clearance rates and even clearance rates for sexual assaults will also tend to be much higher than in the community at large. If the school-based officer(s) are not included in the investigative process, the clearance rate will normally be more in line with even then somewhat higher than what is experienced in the community.

Theft cases in the school setting will typically pose more of a challenge. As in the community at large, clearance rates will often be very low especially in cases of petty theft. The most effective utilization of police resources in these cases is normally in the area of crime prevention education, victim support, electronic surveillance and the use of invisible "snare" powders to identify perpetrators.

A third common category of investigations relating to student violators involves possession and distribution of contraband. Because police in the schools can often base searches for drugs on reasonable suspicion, the problem of drug possession and distribution on campus is normally dealt with more effectively through patrol efforts than through investigation. Because evidence can be located with fewer constraints and dealers can be expelled and banned from the environment, the problem can normally be dealt with without the lengthy and expensive drug investigations so common in the community. Most drug squads arrest the same violators over and over and cannot exclude these individuals from a jurisdiction as a school can.

A common type of contraband investigation will involve a student who is caught with a firearm on campus. In far too many jurisdictions, the student is expelled and sometimes arrested with no more action taken. It is important that investigative follow-up be triggered each time a student is caught with a gun on or near campus. This follow up should include efforts in several key areas:

- A firearms trace through the Bureau of Alcohol, Tobacco and Firearms (ATF) which may link the gun to a theft. This may also help to trigger an investigation if the trace indicates that the gun may have been sold to the student through a gun-running operation. Contrary to popular belief, an exhaustive study of 17 communities completed by the ATF in 1997 showed that crime guns (guns traced in relation to a crime) are normally purchased in the state where they are recovered. Research by the ATF also shows that many crime guns reach the streets through "straw purchasers." If a trace form is not submitted, the ATF will not make the connection between the gun and the gun-runner. In one recent case, two individuals were connected to illegal sales of more than 700 firearms. Not tracing a firearm seized at school may allow the gun running operation that helped the gun get to the campus put more guns into the hands of students.
- All handguns seized should be forwarded to the appropriate crime lab for projectile and cartridge case testing. There are computer programs that allow fired projectiles and cartridge cases to be photographed and loaded into a system that seeks to match samples against projectiles and cartridges that have been found at shooting scenes elsewhere in the region. Because guns can show up in areas far away from the crime location, guns with important value as evidence can be located through this step. The gun that

has been confiscated at school could well be important evidence in an unsolved murder in another part of the country.

- Another important follow up measure is to attempt to search the student's room at home for additional evidence and, more importantly, additional firearms. In most cases, careful persuasion can ensure that parents give lawful consent to search the student's room. In some cases, officers may choose to commit to parents that the student will not be charged for any additional possession charges for guns found in the room. If this is done, it is important that parents be clearly told that the promise concerns possession charges only and specifically does not include charges for any crime that may be linked to the gun. It is important to meet with a representative of the local prosecutor's office prior to this situation arising. Most prosecutors will agree that it is better to get the gun off the street and not prosecute the violator rather than not getting the gun and still not being able to prosecute. Normally, the school can impose school disciplinary sanctions if a gun is discovered during the home search. Another method is to obtain a "gun warrant" to search the student's room for additional guns, ammunition and other firearms accessories to develop a criminal case against the youth. The author believes that this type of search warrant is currently underutilized in many jurisdictions.
- All students who are found with a gun on campus should be asked to tell investigators where the gun was obtained and where other illegal firearms may be located. This tactic can often result in additional guns being taken off of the street.

Investigators should check to see if the student gun violator resides in federally subsidized housing. If so, officers should notify the housing authority regarding the arrest. In some areas of the country, the student and entire family will be evicted from federally funded housing for this type of violation. This may sound harsh but it works as an effective method to deter violators.

Investigations relating to on campus crimes committed by non-students will often involve robbery, assault, theft and distribution of illegal drugs. Often, the violator will be involved in similar crimes that take place off campus. It is therefore very important that local investigators be made aware of these violations. This will increase the chances that the violator will be caught when an off-campus crime is cleared.

Again, a good relationship between students and the school police officer can enhance this type of investigation. Because students will often be aware of information that may be of importance to investigators, it is important that they be predisposed to come forward.

School safety zone investigations will normally be conducted by investigators who are not assigned to work in the school setting. School-based officers should attempt to gain information that may be of use to these investigators and forward it to them. Because many of these investigations will involve assaults in the neighborhood around the school, quick and efficient resolution of such cases is desirable.

Once again, students will often have significant information, and school-based officers should pursue all leads. Often, investigators will have difficulty in obtaining information that school-based officers can obtain easily from a student.

One challenging situation has been the investigation of mass school shooting incidents where there are literally hundreds of witnesses to be interviewed and numerous leads to be following. Following the Heritage High School shooting in the spring of 1999, the Federal Bureau of Investigation (FBI) assisted local and state agencies in this difficult task. The FBI utilizes computer programs that assist investigators keep track of interviews and coordinate the investigation of leads. In the first two days the investigation team interviewed more than 200 witnesses. Assistance provided by the FBI eased the burden on the investigation team and improved overall efficiency of the investigation. Agencies must be prepared to offer full support and cooperation if a successful investigation is to be conducted. There is no place for egos and turf protection in police work - especially in this type of critical investigation.

The last category of school-related investigation involves allegations of criminal conduct on the part of school system employees. These most often involve allegations of physical assault against students, theft of property and crimes involving sexual exploitation of students. It is important that such allegations be carefully and professionally investigated not only to rid the school system of unfit employees, but to clear the reputations of staff members who are falsely accused.

One of the more common types of complaints against school staff is the allegation that some form of unlawful physical force was used against a student. These typically involve a situation in which the staff member has attempted to control a disorderly and, in many cases, physically violent student. It is also somewhat common for a parent to request that a school administrator be arrested for child abuse after the administrator uses corporal punishment. While allegations involving unlawful force by school staff are often without foundation, investigators must remember that these complaints are sometimes true.

It is recommended such investigations be independently conducted by officers who have no connection to the school. If the investigation finds that the facts do not support probable cause for arrest, the use of an independent investigator can avoid allegations of a cover-up by the school and its officers. In those cases where an investigation results in the arrest of a school employee, the use of an outside investigator can prevent a school-based officer from losing effectiveness because of hostility that might arise of school staff believe their fellow staff member has been "railroaded" by police. Given the large number of false allegations against school staff by students or parents seeking revenge against staff members for simply trying to maintain discipline such a reaction is not uncommon.

Theft of property by school staff does not ordinarily generate the strong emotions as other types of investigations. Thefts usually involve musical instruments, computers, cash or other valuables. Often these investigations involve a series of thefts that appear to be connected. Officers who are well-trained in advanced interview techniques will have a much better rate of clearance when working these cases. Advanced training in statement analysis can also be very useful. Excellent training in these areas is available at no cost (tuition, lodging, and meals all covered) through the Regional Counterdrug Training Academy in Meridian, Mississippi. Information on these training programs can be obtained by calling (601) 679-2063 or /3226 or/3227 or/3248.

A frequently effective investigative technique is to run the names of suspected employees against area pawn shop pawn files. In many areas these files are computerized, allowing comparisons to be made quickly and easily which can help identify the perpetrator.

Another effective tactic is to utilize an electronic or chemical stake out. Electronic surveillance using relocated camera equipment can be effective if the equipment can provide continuous taping. Camera location and quality of equipment are critical. There can also be legal issues to be considered regarding the right to privacy in some cases. Chemical stake-outs involve the use of invisible thief-detection powders. One type results in a brilliant purple stain when the powder contacts oils on the suspect's hands. These powders are extremely useful for thefts involving money. The suspect will normally realize what has happened and try without success to wash off the stains yet small spots will remain visible for several days. Often, the guilty employee will call in sick on the following day if they have not yet been confronted. The second type of powder is visible only under ultraviolet light. This type of powder offers the advantage of not alerting the suspect.

The most difficult and critical type of staff investigations are those that involve the sexual exploitation of students. Unfortunately false allegations are not statistically common. The pedophile or hebophile above all else seeks access to children. The result of this desire is that any field or volunteer work that involves regular association with children will attract them. Another trend is that these individuals are often considered to be "good at their jobs" because they relate well to children. This is true because these individuals often have an ability to think along the same lines as the children they victimize.

Often, investigations of this type will be conducted by investigators who work outside school environment since sexual exploitation normally occurs away from the campus. Skilled interviewing by officers with advanced training in this type of investigation is vital. The execution of search warrants focused on photographs, documents, videos and, especially, computer disks and data stored on the hard drive are also frequently utilized. The ideal investigative strategy involves coordinated interviews between the various agencies that have responsibilities, such as the police, prosecutor's office and social service agency which should reduce the number of interviews needed. Investigators must be alert to the alarming possibility of additional victims. This type of offender frequently exploits multiple individuals and in some instances, the number of victims over a period of time reaches triple digit figures.

The Office of Juvenile Justice and Delinquency Prevention of the United States Department of Justice offers excellent training for police officers, prosecutors and social workers in this area. The OJJDP Child Abuse and Exploitation training program is probably the best of its type in the nation. The OJJDP provides this training covering tuition and hotel expenses at various sites around the country. For information about OJJDP training, call Fox Valley Technical College, Criminal Justice Department/OJJDP at 1-800-648-4966.

Chapter 8

SAFE SCHOOL PLANNING AND CRITICAL INCIDENT MANAGEMENT

The topics of planning for a safe school and of planning for those emergencies that may arise are closely related but separate topics that have significant importance in every school. As example after example illustrates, hazards can present themselves in any school setting and though critical incidents are not statistically common, they occur with enough frequency to be a concern. In those schools that are unprepared, the lack of preparation can cause misery beyond description and civil liability, sometimes of staggering proportions.

For purposes of discussion we shall use the term "safe school plan" to address the document that articulates what is being done to prevent tragedy. For situations that involve planning for, responding to, and recovering from emergencies that occur, we will use the term "emergency operations plan". While the particular name that you chose for each may be different, it is important to have a term that illustrates the differences between them.

Safe school planning primarily involves careful planning and design to create an environment that is reasonably safe. This process by definition also requires that steps be taken to prevent injuries and deaths caused by accidents or intentional actions. The safe school plan is a written document that outlines what measures are in place to prevent tragedy. The safe school plan should include all prevention efforts, even though individually some may seem to be mundane or even unimportant. For example, a safe school plan should include educational programs, mental health efforts, physical security measures and any other concepts that are part of a comprehensive strategy to reduce danger. The safe school plan should contain every effort that is made to create a safer learning environment.

It is recommended that a community-based committee or task force be utilized to develop the safe school plan, involving a variety of individuals representing the different disciplines and different segments of the community. Various public safety agencies, mental health providers, government officials, students, parents and others can all contribute to the development of an effective plan. The plan should be reviewed and revised each year and copies of each year's plan should be kept on file for use in such instances as grant applications and litigation following an incident.

Emergency operations planning is best described by the adage "those who fail to plan, plan to fail". A properly developed and practiced emergency operations plan takes over when critical incident management-the actual process of managing a crisis once it occurs-begins.

The first step in developing an effective safe school plan is to conduct a needs assessment for each facility. The needs assessment will be most accurate if it involves work by a group or committee that will draw on the viewpoints of public safety personnel and emergency management personnel, administrators, faculty members, students and parents. The various perspectives that will be offered by these divergent groups can help to ensure that important areas are not overlooked and that plans are realistic.

It is important that adequate information be available to all committee members. They must be made aware of incidents that have taken place if they are to develop a plan to prevent a recurrence and to build a base of support for adequate safety measures. Unfortunately, there are many cases where effective safe school planning does not take place because of the tendency to downplay or even cover up existing problems.

The safe school committee should review certain key areas during the needs assessment that would include:

- incident types
- incident rates (are they going up, down, or are they stable?)
- indicators that gang activity is present
- increased occurrence of graffiti
- crime rate off campus in close proximity to the school
- homicide rate for the community (is the rate rising, stable, or declining; how many school-aged offenders and victims?)
- changes in the truancy rate (some truancy can be related to children who skip school because of fear of being victimized at school or while en route)
- changes in the rate of disciplinary actions
- trends relating to the number and types of weapons being recovered from students
- trends relating to the number and types of drug/alcohol violations
- screening of school system personnel (do criminal history checks, reference checks and drug screening weed out unfit employees and those more likely to commit acts of workplace violence?)
- indicators of cult or extremist groups active in the community (hate groups, militias, satanic cults, and etc.)
- restriction of visitors' access
- familiarity of all school employees with emergency procedures

- current emergency telephone directories in key locations
- availability of updated site surveys
- mock emergency exercises
- possibilities of tornadoes, earthquakes, tropical storms, lightning strikes or flooding
- adequate burglar alarm coverage
- after-hour security of doors and windows
- photo i.d. cards
- student dress code (is it easy to show gang affiliation and hide weapons?)
- trespassers on campus
- identification of trouble spots
- security cameras (do they provide adequate coverage?)
- plans that meet counseling needs of students and staff in the event of a crisis
- dealing with the media in the event of an incident
- dealing with parents in the event of a crisis
- traffic accidents
- traffic violations that can affect school safety
- signs of drug and/or gang activity
- bomb threats
- frequency of fights and other acts of interpersonal violence that can escalate into more serious acts of violence
- adequate responses to children who are being teased or tormented by other students
- treatment of tips that a student may have a weapon
- students' ability to report those who may have drugs or weapons
- procedures specifying how students will be checked for weapons when information is received indicating that a particular student has a weapon
- metal detection equipment (if not routinely used, is it available in the event that a heightened threat level might require it?)
- clear notice regarding school rules to students and parents
- prosecution of students who are caught with drugs and weapons
- knowledge of students concerning what types of conduct will result in criminal charges being filed
- awareness of school officials and emergency service providers concerning contingency plans
- handling of volatile parents who come to the school for a conference
- open or closed campus
- parents who sometimes sign their children out of school early due to fear of an incident
- racial tensions
- recent redistricting that may put students from different neighborhoods

(and different gangs) together
- educational programs that deter violations and teach students how to deal with specific types of threats
- sudden threat to a staff member by a significant other who has become dangerous (could escalate into an act of workplace violence)
- dealing with a mentally disturbed and violent individual on campus
- legally banning from the property those individuals who might pose a threat
- banning suspended students from the facility during the suspension period to prevent problems (are they also banned from school activities?)
- banning from their child's school non-custodial and custodial parents who have demonstrated a danger
- child custody disputes that occur on school property
- adequate policies concerning school searches, sharing of information between the police and school officials and the arrest of students for on-campus misconduct
- training of staff members in recognizing indicators of gang and cult activity, disguised weapons, street drugs, along with proper techniques to use in breaking up a fight
- alternate means of communication if the telephone system becomes inoperable
- additional specific needs of individual facilities

Many facilities may consider many of the above topics so unlikely that they are prone to omit them from the safe school plan. It is critical that all possible issues be addressed. While urban school systems in communities with a high crime rate may have a higher threat level, small schools in many cases experience tragedies because they have not prepared for avoidable threats as well as their urban counterparts.

The safety team should deal first with known and obvious threats. Trends that indicate the likelihood of potential problems that have not been previously encountered should be addressed next. If the facility has never experienced a serious weapons assault even though increasing numbers of weapons assaults are taking place in the community, the school should be prepared for such a possibility. This is particularly true if students are involved with these off-campus incidents or if gang activity is involved. Such situations frequently move into the school setting if they are present in the community.

The committee should review what is being done in other school systems to deal with these problems. The committee should also research the area of school safety planning from news articles that are published regularly on this topic. While always considering proven standard tactics, the committee should also try to be innovative. This will provide the benefit of the "tried and true" and the advantages of innovation. The committee should keep in mind that it is always best to customize a plan rather than simply to copy another school system's plan that may fail to address critical issues unique to a particular facility.

Even though individual strategies are typically found to be in place to prevent particularly incidents, these components are often not viewed as pieces of a larger strategy. Considering these will not only highlight areas where improvement is desired, but will measure the effectiveness of those efforts already in place. This method also helps to ensure that the various components are properly coordinated so they compliment rather than compete with one another.

One possibility is to utilize government school safety experts or private consultants to develop a safe school plan. Caution should always be exercised, however, when dealing with consultants and "experts." Tragically, the explosion of people who offer their services to schools and police includes many individuals who are limited in the quality of assistance that they can provide. Some offer advice that is even dangerous.

Fortunately, there are a number of skilled and qualified experts in both the public and private sector. When seeking outside expertise, it is advisable to find out:

- How long they have been delivering these services.
- What type of training they have attended.
- What educational credentials they possess.
- What type of practical experience they have had.
- What several different references have to say about the quality of their services.

It is also important to consider if the consultants intend to assist the local community in developing a quality safety team, or if they intend to make you reliant on the repeated use of their services.

Also, you should ask if they conduct their own research and development, or do they simply steal ideas from other sources which leaves them with a limited ability to expound on those concepts. There are a number of individuals who conduct business in this manner and they are typically unable to deliver quality service. Beware of the expert who uses the information of others without giving proper credit to the actual source. This is a good sign that they are probably not capable of helping you at the level that you should expect and deserve. These individuals can also expose the contracting organization to civil actions relating to unauthorized duplication of materials that the consultant has no right to allow the organization to use.

In many cases, a little legwork can reveal that government agencies provide free or very inexpensive technical assistance, training and consulting services that are of comparable or higher quality than that of the for-fee providers.

Whether free or for-fee, it always pays to evaluate the quality of the service provider. Just because an agency is government-funded or is expensive, does not mean that it can provide quality services.

The Georgia Emergency Management Agency, Office of the Governor, is mandated by law to provide free school safety technical assistance and training to public and private schools in Georgia. The Agency provides a wide range of services in the areas of school safety, crisis planning and crisis response capabilities. GEMA makes information freely available to those in other states through its school safety web site at www.state.ga.us/GEMA/. GEMA staff also offers limited assistance to practitioners from around the world. Be sure to check with government agencies in your state to see what support is available before hiring consultants to provide services that may be available to you at no cost.

Each of the areas listed in the assessment phase should be addressed. The team should make a list of recommendations for safety measures that are deemed appropriate to the facility. In a small rural school where an outsider would be quickly spotted and most people know each other on a first name basis, student identification cards may serve little purpose. In the same school however, failure of staff members to intervene in a situation where a student is repeatedly being teased by fellow students might result in a serious weapons incident more quickly than in a large urban school setting.

Some specific suggestions for improving school safety might include:

- Develop a standard reporting criteria that requires that all criminal acts on campus be reported to the school resource officer.
- Design a standardized means of tabulating all criminal incidents; a computerized records management system can be used to maintain these records more efficiently. The ultimate means for tracking crime on and near campus is to use a computer mapping system that is compatible to a computerized records management system.
- A copy of the current bomb threat procedure should be on hand at all locations where bomb threats are likely to be received. The plan should be reviewed periodically to see that it remains viable.
- Ensure that students and staff are evacuated at least 1000 feet from affected areas during bomb threats if evacuation is deemed to be the safest course of action.
- Select multiple evacuation sites for bomb threats and switch sites frequently to reduce the dangers from secondary explosives being planted in evacuation areas.
- Have personnel designated and trained to check evacuation areas for suspicious packages/items during bomb threats.
- Have armed police personnel present on the perimeter whenever students are evacuated (even during drills) as a deterrent to a violator who may plan

to shoot at students upon evacuation. Just as importantly, this will help to ease the fears of students who may be afraid to evacuate the building.

- Have armed police personnel conduct periodic surprise patrols of wooded areas near schools as a deterrent to a potential sniper. Doing this in a highly visible manner not only lets people know that this precaution is being taken as a deterrent and but tends to give students and staff peace of mind.
- Conduct annual reviews of statistical data to note changes in incident trends. Larger systems may require more frequent reviews.
- Ensure that all incident reports that involve gang activity be properly marked "gang-related incident" and forwarded to the local law enforcement gang unit for intelligence purposes.
- Photograph gang graffiti on school property before removal. If large amounts of graffiti are found early in the morning, students should be kept away from that area until it can be removed. In many cases, fights may break out between members of the offended gang and members of the gang that placed the graffiti.
- Review regularly statistical data relating to serious crimes that occur during school hours in the school safety zones around each facility. School safety zone enforcement should then be coordinated to target areas that cause the most concern.
- Increase preventive weapons screening measures following major incidents of gang violence.
- When truancy rates increase for a school, truant students should be interviewed to determine how the increase may be related to fear of victimization at the school or in the school safety zone.
- Be alert for significant changes in the rate of disciplinary actions that indicate school safety trends that should be addressed. A sharp increase in fights may indicate increased gang activity and always indicates a greater likelihood of weapons incidents.
- Increases in the number and sophistication of weapons recovered from students indicate that preventive search activities must be enhanced to prevent an increase in the chances of having a serious weapons incident. They are also often connected to an increase in gang activity. Gang activity greatly increases the chances that weapons will be actually used.
- Increases in the number of drug and alcohol violations can also be connected to gang activity. They may indicate that new preventive measures be utilized. These could include the use of drug dogs to check lockers, vehicles and public areas, the use of surveillance and in some rare cases the use of undercover police personnel.
- Criminal history checks, pre-employment drug screening, integrity testing and more thorough reference checks prior to employment of school system

staff will reduce instances of employee theft, crimes involving the sexual exploitation of children and the possibility of acts of workplace violence.

- Staff members can be given training relating to gang and cult activity, extremist groups, dealing with difficult individuals, classroom management, crime prevention, recognition of disguised weapons and street drugs, crisis response and the role of individual staff members in the emergency operations plan.
- Visitor access can be restricted through announcements sent to parents, posted signs and maps, visitor badges (color coded for specific dates or those that fade blank after a specific time period) and by locking remote access doors.
- School employees should receive periodic briefings regarding the school safety plan to ensure that all staff members are aware of their role in it.
- All possible emergency telephone numbers should be readily available. The listings should be updated prior to the start of the school year and should be located in several key locations at the facility.
- School staff should have the authority to call emergency personnel in the event of an emergency. In many cases, precious time has been lost because the building principal had to be located before emergency personnel could be called.
- Tactical site surveys should be completed for all buildings; one copy should be accessible in the main office with another copy at a remote location. Each emergency evacuation kit should contain a copy of the latest site survey. The site surveys should be updated annually. Photographs of key areas inside and outside of the building can be of great use during a crisis. A short video tour of the building can also be an asset to emergency personnel in such cases as a hostage situation.
- School officials should meet with emergency service providers and emergency management personnel to ensure that everyone understands issues that are critical for each in dealing with incidents on campus. Contingency plans can be formulated and discussed before an incident takes place to reduce mistakes.
- Each facility should conduct periodic mock exercises on staff planning days or other work days when students are not present. Emergency personnel should be invited to participate from time to time. Presence of fire service personnel or a police tactical team can be very insightful. It will also ensure that some of the same experts that would be responding to the facility in a crisis are already familiar with the structure. Seek guidance from local and state emergency management agencies to develop exercises. Tabletop exercises are particularly useful in testing emergency operations plans.

- Offer local emergency personnel the use of the facility for training purposes when it is not occupied. Encouraging the S.W.A.T. team, canine unit or fire department rescue unit to use the building will improve the relationship between these agencies and the school as well as help to make emergency personnel more familiar with the structure. It should also help to deter vandalism and burglary (nobody wants the S.W.A.T. team to show up while committing a burglary!). An understanding should be reached to ensure that the facility suffers no damage and that emergency personnel will clean up after the training is completed. It is also a good idea to encourage these agencies to use loaned school buses for training so they are familiar with bus design.
- Ensure that critical incident planning provides comprehensive preparations for those regional natural disasters that may be more likely to involve the facility.
- Fix weak spots in the alarm system coverage and physical security of the building. It is important to remember that a burglar will strike at the weakest point and the level of security is determined by that point.
- Consider the use of the "blackout concept". Studies have shown that contrary to popular belief, schools that are completely without light at night are far less likely to be burglarized and vandalized than schools that are well lighted. For the blackout concept to work, every single light must be extinguished. Of course, the blackout strategy is not to be used during periods when the building is occupied.
- Implement a system to see that the last person to leave the facility each day checks to ensure that all doors and windows are locked. A high percentage of school burglaries involve entry to the building through an unlocked door or window.
- Ensure that all valuable school equipment is clearly engraved with the school name and that all serial numbers are recorded.
- Computers, television sets and other electronic equipment can be secured with cables and other devices that are designed to make theft more difficult.
- Expensive equipment can be moved into more secure rooms for storage during the summer or during holiday breaks.
- Student and employee photo identification cards can increase the level of safety when trespassing is a problem.
- People who have no legitimate business on the property can be warned in writing to stay off school property. This use of criminal trespass warning slips or "barring slips" can reduce the incidence of trespassing. The slips prove that the individuals have been warned and can be used as evidence in court when violators are arrested after returning to the campus. Computerizing such a list is helpful, and photographing individuals who

have been warned may also be useful.

- Although often a controversial approach, the use of school uniforms help to prevent trespassing, neutralize gang identifiers and can help to reduce the rate of student handgun violations. If uniforms are not practical, a strict dress code that requires that shirts be tucked in can be very helpful in limiting where a student can hide a gun. Lack of a proper dress code invites gun violations.
- Trouble spots can be identified with simple pin maps or computer-mapping programs and these areas then targeted with specific strategies to reduce the specific crime problems encountered. Most school administrators can pinpoint the obvious trouble spots that should be a priority for police patrol coverage and monitoring by school staff.
- Incident locations should be coordinated with security camera coverage. In many cases, trouble spots may shift to areas without camera coverage or where image clarity is not adequate to address the problems within the camera's theoretical coverage area.
- A crisis counseling plan that addresses the needs of students, staff and parents should be incorporated into the safe school plan.
- Procedures should be outlined to cover the extent of information that should be given to the media, when it will be released, who will serve as spokesperson and/or liaison, and when (or if) the media will be allowed on the property.
- Specific locations where traffic accidents or violations are commonplace on or near the campus should be identified and targeted for appropriate traffic enforcement.
- If drug or gang activity is common near the school during school hours, specific trouble spots should be targeted for intensive enforcement.
- Call-tracing features should be placed on telephones that are likely to be called in the event of a bomb threat.
- Pay telephones and school telephones available to students should be restricted or removed.
- When a caller is identified and arrested for making a bomb threat, the principal may wish to make an announcement about the arrest over the public address system. No mention of trace equipment should ever be made. Generally, the name of the arrestee should not be broadcast to deter copycat violations common in bomb threat violations.
- One strategy to reduce fights is to have a strict policy that requires students in a major fight to be arrested by citation or custodial arrest. The use of fines has consistently proven to be effective in reducing fights. Remember, more fights mean more weapons, and the combination of the two means a greater likelihood of a serious weapons incident on campus.

- Staff members should be urged to understand that some of our nation's worst school weapons incidents and hostage situations have been at least partly triggered by students repeatedly teasing and/or harassing another student. Staff members can reduce the chances of such an incident by intervening when they become aware that a child is being targeted on a regular basis or in a severe fashion by classmates.
- A procedure should be established to deal with information that is received indicating that a student may have a weapon. This procedure should always involve armed police presence when the student is approached. Numerous school shootings have resulted because civilian administrators attempted to handle these situations without police involvement.
- A procedure should be developed to deal with non-students on campus who are reported to have a weapon. Again, the presence of an armed police officer is a must.
- Students should be well aware that the school will respond to information concerning the presence of drugs or weapons on campus. The school police department telephone number or other points of contact should be posted in prominent places.
- If entry-point screening is not used at the facility, it is a good idea to have such equipment available. Personnel should practice using it from time to time and be familiar with its use. A policy should be in place that allows for its use when required by an emergency. Time after time, a school will find that a new threat requires use of this type of equipment only to learn that the equipment cannot be purchased and put into use in time to be of any value.
- Posters and signs in prominent locations that advise students, parents and visitors that weapons are not allowed on campus should point out some of the search strategies in use and the consequences of being caught with a weapon on campus.
- A clearly written student handbook should be used district-wide to provide standardized guidelines. This helps to avoid confusion when students transfer from one school to another. All students must receive a copy.
- A contract may be utilized where students, parents and the school agree to work together to keep weapons and violence from being tolerated on campus.
- A protocol should be worked out to deal with parents who exhibit a potential for disruptive behavior or acts of violence coming to the school for conferences. One strategy is to have a school resource officer present when such meetings take place.
- Measures can be taken to restrict access to an open campus when problems begin to arise that indicate that the campus is too open.

- Growing racial tensions on a campus may indicate that efforts need to be made to incorporate cultural diversity educational programs into the curriculum.
- A policy prohibiting racial discrimination and hate crimes can ease tensions between different groups on campus.
- Redistricting may require that bus schedules be reviewed to reduce conflicts on buses where an unhealthy mix of students from rival gangs can occur after the changes are implemented.
- Educational programs taught by school resource officers should be focused to address problem areas identified by the school safety team.
- Staff members should be asked to inform the school administrator if they experience threats from a former spouse or significant other. Key office staff members and the school resource officer should then be shown photographs of the individual and a procedure worked out for appropriate response if the individual comes to the school. If the individual is reported to have access to or be known to carry weapons, the school resource officer may need to conduct a pat-down of the individual and search the individual's car. It may also even be appropriate to ban the individual from the campus. It is also a good idea to have an officer detailed in the area if the person is expected to appear, particularly at an elementary school.
- School staff should be advised to report immediately to the school resource officer any suspicious persons and mentally disturbed individuals on campus.
- School administrators may be issued portable radios to allow fast reporting to the police officer on duty at the school. A portable radio can greatly enhance the safety administrators when they find themselves involved in a tense situation.
- Radios can be available for teachers when they are taking students outside. These can be particularly useful when the group will be in a remote area of the campus.
- Students who have been suspended from school may be banned from the campus (and school safety zones) during the suspension period. This will reduce disruption being caused by suspended students coming onto the campus.
- In most states, non-custodial parents and even custodial parents can be banned from their child's school if they have shown that they are likely to cause disruption or disregard school rules for visitation, or if there is an indication that they pose any sort of danger. The building administrator should always be involved in any decision to ban a custodial parent from the school. Of course, such action should never be taken without careful consideration.

- Citations and, if needed, custodial arrest can be an effective means to deal with the parent or other relative who disrupts the school or improperly boards a school bus. Allowing parents to curse and yell on school property sets a tone where acts of violence are more likely to occur.
- A procedure should be worked out regarding a parent(s) who comes to the school to take custody of a child when that parent is not listed as authorized to pick up the child. Great care should be taken when being asked to relinquish a child because of court-ordered custody. In fact, there may be statutory law in place that does no allow custody to be transferred at school. A police officer who takes custody of a child pursuant to a court order may be in violation of a criminal statute. This matter should be researched beforehand and a protocol developed with the local police agencies.
- School system policies should be reviewed to ensure the address of such critical areas as school search and seizure, sharing of information with police agencies, arrest of students and dealing with students who are arrested off-campus for crimes that have an impact on school safety.
- Teachers and other staff members can be provided with training to help them handle fights and other disruptive behaviors.
- An alternate means of communication should be established for the facility. The use of portable radios is often a viable option. Cellular telephones are another possibility. All staff should receive training on the safe use of such equipment (for example, a cellular telephone or radio can possibly set off explosive devices).
- Parent volunteers may serve as hall and parking lot monitors. Volunteers may also be utilized for special events such as school dances, athletic events, spring festivals and the like. Volunteers have been successfully utilized to secure doorways, inspect baggage, operate metal detection checkpoints and perform a host of other duties. Of course, background checks must be conducted on all volunteers.
- Restricting access to vehicles by students may reduce their opportunities to retrieve weapons, drugs or alcohol.
- An evaluation of each facility by a person trained in crime prevention through environmental design (CPTED) should be conducted. Plans for new facilities should also be reviewed in this manner.

A key component of safe school planning is the Tactical Site Survey. The survey can help to increase the odds that vulnerable locations in the facility will be identified. More importantly, the properly completed site survey will be valuable to emergency personnel during a crisis. It is a critical point that the complete site survey report be on hand not only at the school office, but at a remote location (i.e. local emergency management office) in case the office copy is not accessible during a crisis.

The tactical site survey should be recorded on a site survey form to ensure that all key areas are addressed. Each survey should include:

- Name of the person(s) completing the survey.
- Date the survey was completed or last updated.
- Location of surveyed facility with street address.
- Number and types of floors.
- Approximate square footage.
- Exterior construction type and type of roof.
- Number of students, faculty and staff normally in the building.
- Type of fire sprinkler system and location of sprinkler controls; external fire department connection.
- Access to adjoining rooms through ceiling spaces and/or air ducts.
- Weight that roof will support for possible helicopter use and clearance for helicopter rotor blades.
- Helicopter landing coordinates for several nearby sites.
- Coordinates for the center of the campus to enable Federal Aviation Administration to designate the area as closed airspace during any crisis (to help keep media helicopters from hovering).
- Intersections (at least 1000 feet away) that need to be blocked with such information to be programmed into any police department computer-aided dispatch system and school safety zone signs be placed at each site for a landmark for responding police units.
- Underground or elevated connection(s) to an adjacent building?
- Type of perimeter fencing.
- Emergency evacuation plan that is clearly posted (attach copy to site survey).
- Description of internal intrusion alarms with location of system schematics.
- Fire alarm system(s).
- Description of closed-circuit television system providing inside or outside coverage with location of cameras and monitors and shut-off controls (so the violator cannot use the camera system).
- Audio monitoring system (interior or exterior) with locations of sensors and monitoring devices.
- Number of assigned police personnel with telephone number of the campus police department.
- Number of assigned security personnel.
- Principal's name and home telephone number.
- Assistant principal(s) name(s) and home telephone number.
- Head custodian's name and home telephone number.
- Location of duplicate building keys.
- Maintenance department telephone number.

- Location of central power control panel.
- Location of building plans (contact and telephone number, if necessary).
- Police department telephone number.
- Sheriff's department telephone number.
- Fire department telephone number.
- Emergency medical service telephone number.
- Local emergency management agency telephone number.
- Hospital telephone number.
- Address of nearest hospital.
- Pediatric emergency capacity, hazardous material emergency capacity, etc. of local emergency medical facility.
- Phone numbers of other hospitals in the region.
- Driving route to nearest hospital.
- List of students and staff members with medical conditions that could be critical in an extended hostage situation or if these persons where trapped in a section of the building during a natural disaster.
- On-site parking area for large numbers of emergency vehicles.
- Site staging area located off of but nearby the site.
- Nearest off-property helicopter landing site.
- Other tactical concerns.
- A photo album containing photos of all key areas

The emergency operations plan should be developed through a community-based group and designed to fit that particular community. An important caution here is to avoid the "plan-in-a-can". Plans that are developed simply by taking an existing plan and making minor changes can prove to be disastrous in the event of a major crisis. In addition, reliance of popular flip charts as a stand-alone plan will typically prove to be inadequate in the event of any major incident.

A critical point is that all area public safety agencies and mental health providers who may have to respond to a school crisis be involved in the development of the emergency operations plan. This will help to ensure that the plan is realistic and that everyone clearly understands their role. It is easy to underestimate the actual difficulty of a school crisis situation. Bringing local experts "to the table" can help to reduce oversights that might otherwise cripple response and recovery efforts.

An effective concept is to develop a district-wide plan or protocol to establish consistent standards for every facility in the district. This document will ensure that necessary steps are taken no matter where the crisis occurs. Once this document is completed, specific procedures should be developed for each type of incident for each facility. These procedures are then inserted into the appropriate section of the protocol. This method helps to make sure that there is an adequate and consistent set of guidelines in place for each facility while also making certain that the overall protocol is tailored to the unique needs of each facility.

As an example of this system, consider evacuation for a bomb threat. If it is determined that the situation mandates evacuation, the protocol might require that students be evacuated to an area that has been swept by a "go-ahead" team prior to the arrival of students, that multiple sites are also available and that the evacuation site is at least 1,000 feet from the school. The individual procedure for such an incident might specify which sites are available for the school, who shall serve on the go-ahead team and what routes are designated for reaching each potential evacuation sites.

There are a variety of CDROM programs that allow plans, schematics, photographs and even video to be stored on disk. There are also computer-based critical incident management programs commercially available. Flip charts can also support the plan during the initial minutes of a crisis. It is important that these supplements are designed from the district-wide plan. In some cases, local agencies or parent volunteers have helped to develop excellent supplements of these types.

Emergency operations plans must address potential issues such as:

- The point of contact for the media during a crisis: It is important that this function be centralized to avoid the releasing of conflicting information or information that could increase danger (this is particularly critical during hostage situations when improper release of information can cause the death of hostages or tactical personnel).
- Emergency chain of command: This should provide clarity regarding who has authority and responsibility for various emergency duties. This should be coordinated with local public safety and emergency management agencies in advance.
- Incident command structure: Typically, the local public safety community has such an established structure. The school emergency operations plan should incorporate this structure.
- Records: A system should be established to record major actions, state who took each action and denote the time each action was taken.
- Evacuation and lock-down codes and procedures: It is normally best to avoid the use of complicated code phrases such as "Mr. Jones has left his blue binder in the building." Such codes often prove confusing. Several levels of lock-down should be developed to fit situations of different threat levels.
- Planning for special events and events involving students who are traveling to and from school by bus should also be included.

- Plans should address incidents that occur off-campus but in close proximity to the school that pose a danger.
- Plans should include considerations for special needs persons. For example, making sure that a system has been developed to evacuate students who must depend on wheelchairs.
- Media staging area: Plans must anticipate the prompt arrival of large numbers of such personnel. In police tactical situations it may be crucial that they not be allowed within sight of the crisis. Media coverage can provide a hostage-taker with the locations of tactical personnel, not only endangering their safety but making the situation more difficult to resolve. It is generally best to provide a media staging area and to provide periodic updates through a designated spokesperson. It is best to select multiple sites for each facility.
- In hostage situations, media should be kept from announcing the presence of key officials such as the school superintendent, mayor, police chief, school board president and similar officials. Including them seriously handicaps the negotiation process, and in some cases has contributed to the death of hostages because hostage-takers often demand to speak to such key officials. Because these officials are considered to have authority to act, hostage-takers may become angry when demands are not met immediately and kill a hostage. Trained negotiators can often "buy" valuable time because they lack the authority to grant many requests. Provisions should be made to keep key officials out of sight in a think tank if they are on the scene. Reporters may be told that officials are unavailable.
- Emergency medical evacuation: In one rural elementary school hostage situation, adequate provisions were not made even though the hostage-takers indicated that they had an explosive device. When the device was accidentally detonated and more than 80 victims were injured, extensive time was required to evacuate them.
- Recording names of all students and staff who leave the incident scene and who they left with: In crisis situations victims may be evacuated to various medical facilities. It is important that all victims be listed along with their destinations. In addition, students who are picked up by relatives must be listed along with the name of the person who picked up the student and the time. Several family members will often attempt to pick up a student. Once the child has left with the first family member to arrive, other family members should be informed of the child's status. Determining exactly which students are still trapped in a facility during a hostage situation or natural disaster can be a life-or-death matter.
- Uninjured students may need to be moved to a designated staging area away from the building, so that they may be picked up when necessary.

This area should facilitate an organized pick up process. A bullhorn can be worth its weight in gold at this location.

- When a major school bus accident takes place, the same type of process should be used as listed above. Such precision is necessary because some parents, unfortunately, may encourage their children to claim injuries on the affected bus in order to file a civil suit against the school system.
- Family members and relatives: Another common problem is the arrival of large numbers of such individuals. Several relatives may arrive for each student. In fact, in most crisis situations, family members, media representatives and curious spectators constitute the bulk of the perimeter security problem. Managing this crowd can pose as many challenges as the management of the crisis situation itself. Quickly establishing a staging area or family reunification area for family members and relatives makes this challenge more manageable. This can be the most challenging and complex aspect of crisis planning and response efforts. Finding a secure site that is remote from the school and is always available can be difficult. This is especially true since literally thousands of parents and relatives can turn out for a school crisis. To ensure the security of the site, it is best not to preannounce its location. Schools should never be used as a family reunification site. Attempts to use another school as a site are likely to result in even more confusion as it results in two schools being in crisis. Announcements by the media (prepared in advance) can advise parents to go to the proper site rather than the school. A team approach to preplanning will cover most major obstacles.
- Each school should have two evacuation kits that contain all items critical to the crisis management process including:
  - the school system crisis protocol and building crisis plan
  - most recent yearbook
  - pencils/pens/paper
  - radio/cell telephone
  - tactical site surveys
  - current student and staff emergency contact cards
  - high visibility vests for administrators
  - roll of quarters for pay telephone calls since cell telephone systems sometimes collapse during a crisis
  - emergency telephone numbers for key personnel.
  - information sheet on the school for the media (number of students, faculty, programs offered etc.).
  - bullhorn
  - other items as desired

The kits should be in two separate locations in the building in case the office is the incident location. Both kits should leave the building whenever the person who is assigned responsibility for each kit evacuates the building for a drill or for an emergency. A small rolling suitcase with collapsible handle makes a good container for an evacuation kit.

Critical incident management preparation should cover any possible crisis situation even if such a possibility seems extremely remote. Some examples of possible crisis would include:

- a major weapons assault
- suicide by a student or employee
- bomb threat
- detonation of an explosive device or an accidental explosion
- detonation of a large vehicle-type explosive device
- power failure
- flood
- civil unrest
- fire
- forest fire
- tornado
- hostage situation
- multiple victim homicide
- barricaded gunman
- serious assault
- aircraft crash
- lightning strike
- chemical/biological hazardous materials incident
- earthquake
- hurricane
- severe hail or snowstorm
- poisoning (intentional of accidental)
- vehicle striking students and/or building
- mass trampling
- dangerous animal loose
- tidal wave
- severe winds
- volcanic eruption
- mudslide
- contamination of cafeteria food or drinking water (intentional or accidental)
- contamination by toxic fumes or irritants such as students spraying pepper spray in the air conditioning system (intentional or accidental)
- weapons of mass destruction

While a school system in Iowa need not plan for a tidal wave, the planning should consider every type of crisis that could possibly occur in the region. Even during a crisis that could not have been anticipated a school with a comprehensive safety plan will often find that the plan provides adequate preparation.

It is also important that the completed plans are not simply placed on a bookshelf to gather dust. Each school system employee should receive regular training on his or her role during a crisis. It is also necessary to test the plans through appropriate exercises. Both table-top exercises and full scale exercises should be coordinated through local and state emergency management agencies to determine flaws in plans before they are tested by a actual crisis situation.

The emergency operations plan should be viewed as a living document, reviewed and revised every year, that must change as the resources in the community change. It must also meet the ever changing threats in our fast paced society.

A good example of a comprehensive school crisis plan saving lives in situations that could not be anticipated occurred in a rural south Georgia school. A student brought what he thought was an inert dummy training round from a military grenade launcher to school. The student's father had found the round 20 years earlier while he was working on a military base. During the school day, numerous students and even staff members handled the item without suspecting any danger. When one of the students stood on a chair and dropped it on the floor, the grenade detonated and because it was an anti-personnel round, it sent shrapnel flying throughout the room.

After the principal heard the sound of the explosion, he activated the school emergency plan for an explosion immediately which called for evacuation of all non-affected students and staff members to a staging area a safe distance from the school and notification of authorities. Approximately a dozen people were injured, several students critically. Following the plan for mass medical evacuation, several hospitals were contacted and emergency medical helicopters were dispatched. All injured personnel were quickly evacuated.

By moving non-affected personnel to the staging area, a major problem was avoided. Typical of these incidents, radio stations monitored the emergency radio traffic and made immediate newscasts, which resulted in numerous relatives converging on the school. Because of proper planning, parents went to the staging area, allowing emergency personnel to move freely in the building without fighting through panic-stricken crowds. The non-affected students were released to relatives in an orderly fashion with care being taken to record who had picked up each student. This prevented the chaotic and time-consuming task of trying to reconstruct where a child was when a second relative arrived after the student had left with the first relative.

Using designated helicopter-landing zones and the efficient manner in which other hospitals were utilized to prevent overloading a small rural facility (fortunately located next to the school) ensured that critical time was not lost. It is clear that several students would have died from blood loss (one student had to be revived after his heart stopped) if an effective crisis management plan not been in place. Obviously, few people would have anticipated an accidental detonation of an anti-personnel grenade in an room full of students. Because a rural school considered the remote possibility of an explosion, however, the plan dealt effectively with this unusual situation.

This example and many other situations that have occurred illustrate the value of comprehensive planning in the area of critical incident management. The National School Safety Center has compiled a great deal of information regarding safe school planning and crisis management. Their personnel can provide exceptional training, consultation and technical assistance to school systems, police agencies and emergency management organizations. The Georgia Emergency Management Agency, Office of the Governor, has also assembled a great deal of information pertaining to these topics. The agency can provide technical assistance and training through its school safety project. The Federal Law Enforcement Training Center in Glynco, Georgia, also offers training programs in the area of critical incident management with an excellent instructors' program relevant for larger school systems.

Adequately addressing all critical areas for safe school planning and crisis preparation requires a great deal of effort, but such steps are blessedly welcome when a crisis is avoided or when one takes place and measures are in place deal with it effectively. Safe school and emergency operations planning are an absolute necessity in today's world. Those schools that do not take adequate steps in this area are playing a dangerous game of chance with the lives of those in the school community as the stakes.

Chapter 9
BASICS OF SCHOOL SEARCH AND SEIZURE

The law relating to school search and seizure and other areas important to police in the school setting has undergone dramatic change over the years. Fortunately, both statutory law and case law has enhanced the authority of school officials and police officers to address those challenges to school safety that have grown more important in recent years.

Individual school systems and school police units should always conduct independent research into laws governing school search and seizure because there is variation from state to state and new case law is being handed down regularly. Methods that have been developed and implemented in the Bibb County Public School System involved legal research by the school board attorney and the Campus Police Department.

The Campus Police Department has relied heavily on the advice of Dr. Gary Avery, an attorney with extensive experience dealing with school/police matters. Dr. Avery has served as an instructor for Law Advisory Group which specializes in providing training in the field of school/police legal matters. Many school board attorneys, prosecutors and police academy instructors are not well versed in this unique field. Attorneys who specialize in these matters and legal groups such as the Law Advisory Group should be consulted prior to major changes in a school's search and seizure policy or its preventive search strategy.

Much of the information in this chapter comes from what the author has learned from seminars taught by Dr. Avery and from studying the manual *Safety, Order and Discipline in American Schools-Defining the Authority of Educators and Law Enforcement* published by the Law Advisory Group. The author has found that these are among the most up to date, reliable and straightforward sources of legal guidance for the school police officer or school administrator. Dr. Avery as a matter of routine will provide brief solid legal advice without charge over the telephone and is a very competent legal instructor. Dr. Avery can be reached at 954-785-1596. Law Advisory Group offers reasonably priced seminars taught by trained attorneys at various sites around the country. They also sell their manual, which is frequently updated. No school police operation should be without a current copy of this manual. Law Advisory Group can be reached at 216-371-0212.

This chapter is designed to provide only the bare basics of school search and seizure. Again, each district should conduct its own legal research to ensure that regional legal points do not pose a problem.

One of the landmark cases in school search and seizure is the famous New Jersey vs. T.L.O. Case, which was decided by the United States Supreme Court. In this case a high school teacher brought a student to Vice Principal Theodore Choplick's office for smoking. Vice Principal Choplick searched the student's purse without additional facts that would constitute probable cause. Basically, the court ruled that school officials can:

- Search students based on the standard of reasonable suspicion.
- Search without having to meet a requirement to obtain a search warrant.
- Question students without advising them of their Miranda rights when questioning them pertaining to a school disciplinary matter even if they may report admissions of a criminal nature that will subsequently be reported to the police.

The T.L.O. case has tremendous significance in several ways. Applying the standard of reasonable suspicion means that school officials can search based on a considerably lower standard than that often required of a police officer.

The T.L.O. case did not specifically address the issue of searches by police officers in a school setting. While a police officer was present when Vice Principal Choplick conducted his search, the officer did not actively participate in the search. The court specifically chose not to address the issue of police searches in schools at that time. Following the T.L.O. decision, it was common practice for school administrators to search students based on the standard of reasonable suspicion. Frequently, police officers and school administrators were instructed that it was best for the school official to conduct a search while the police officer waited outside the office. Once the administrator found contraband, the officer would arrest the student. This practice obviously placed school administrators in substantial danger.

As the danger became more common, the courts became more aware of it, and a California case emerged that addressed this issue in 1982, In re Fred C., a juvenile, (1982) 102 Cal Rptr 682. Although this appellate case is not necessarily binding in other regions, Dr. Avery states that it has been successfully utilized in all 50 states. This case enables school administrators to bestow the school's authority on the school police officer under certain conditions.

Under this ruling the officer can assume the school's enhanced authority to search based on the standard of reasonable suspicion when:

A. The police officer is at or has come to the school at the school's request which means that the school administrator (or school system administration) welcomes the officer's presence. If an officer were to arrive at the school and try to impose his or her authority against the wishes of the administration, it would not meet this standard. Campus Police Departments and School Resource Officer arrangements will automatically meet the requirement under most conditions.

B. The school must have the authority to search in the first place. The school must meet the standard of reasonable suspicion before it can search, and the officer must also meet this requirement.

C. The educator or school system must be in charge. This can be by direct involvement or through policy. For example, the school board may prepare a written policy specifying when school police personnel may assume this expanded authority. The bottom line is that the officer must understand that the expanded authority is given by the school system and can be revoked by the school system at any time.

D. The officer must be able to demonstrate that an officer's participation was required to reduce danger to students and school officials, normally an easy standard to meet. The danger can be from the potential of a variety of situations. Common concerns could include:

- the student attacking the civilian administrator,
- the student attempting to flee the scene and getting hit by a car,
- the student having a weapon,
- the search requires expertise or special equipment such as that required to search a car safely and properly,
- encountering hypodermic needles, razor blades or other dangerous implements.

In searching for a suspected weapon with adequate cause, the officer should first ask the student about possession of any item that is prohibited by law or policy. If he or she fails to disclose such an item, the officer should pat the student down for weapons. If the student is being searched because they are suspected to be carrying a firearm, it is recommended that the front waistline and pockets on the student's strong side be checked first. A systematic pat down of the entire person should then be conducted. In most situations a student with a gun will have the weapon concealed in the waistband or a pocket on the strong side.

In searching for contraband other than weapons it is best from a legal standpoint to utilize the "least intrusive means possible" in conducting the search. Asking the student to empty all pockets and pull clothing tight against the body along with other such measures is advisable before resorting to a hands-on pat down.

It is an important point that particularly careful research in the area of strip searches should be undertaken. This is an area where reasonable suspicion will not suffice. In most states a very high degree of probable cause will be needed to conduct a legal strip search in a school setting. In some cases it may be practical to make a custodial arrest for another charge and then conduct the search. A typical example of this type of situation would occur when the student attempts to fight with officers after they have attempted to conduct a legal search that is more limited in scope.

Another useful legal measure is the extensive use of preannouncement and adequate posting of search policies. It is important that students be notified through posters, metal signs, parking decal applications, video presentations and other such means that individual searches will be conducted based on reasonable suspicion. In the case of random surprise searches, no individual suspicion is necessary. Students should be advised that school lockers are property of the school and subject to search at any time for any reason. It is a key legal point that only school-owned locks be allowed. Parking sticker applications and signs posted at every possible vehicle access point to school property should advise students and visitors that "by entering this property, the driver of any vehicle automatically consents to the search of the vehicle and any containers therein by school officials and/or police officers with or without cause". In most areas this type of preannouncement will allow for legal searches based on less than reasonable suspicion. This type of sign should also make the school undesirable for drug dealers and weapons violators driving a car that contains contraband.

Every means to inform the student body about individual and random weapons searches will further enhance the legal position of the police officer - and, just as importantly, it will reduce weapons violations. Another key issue is the general concept of using random searches as a reasonable measure in light of the current threat level. It is very important to be able to show in a court that the level of danger that is or would be present

It is always a good idea to clearly advise students and visitors to campuses about search policies.

justifies the level of intrusion posed on the student body by the searches. It is critical, therefore, that accurate records be kept relating to past weapons violations and assaults on campus.

The field of school safety has evolved to the point where there is probably more exposure to civil liability from failure to take action than there is from taking reasonable actions. In one of his seminars, Dr. Avery reported that it is not unusual for out-of-court settlements and jury awards to result in the range of $5 to $7 million per violent school death when it can be demonstrated that a reasonable level of safety was not provided. The courts seem to be sending a strong message to school and police officials that changes must occur to make our schools safe again and that those who fail to heed that message should be prepared to pay as dearly as the victims of school crime.

Chapter 10
DEALING WITH YOUTH GANGS, CULTS AND VIOLENT EXTREMIST GROUPS

Gang activity in a community is a serious issue for any school located in that community. If gang, cult or violent extremist group activity is present in the community, it will have an effect on the schools. The presence of one or more of these types of groups dramatically increases the chances of major weapons incidents and other acts of violence. Many of our nation's most nightmarish school incidents have involved students or non-students who are connected to one of these three types of groups.

Youth gangs are not a new problem in our nation. In fact, youth gangs in America actually precede our own government. The earliest recorded references to youth gang activity in America of which the author is aware dates back to the 1760's. During this period, groups of young men began to commit crimes as a group. Normally, their crimes involved highway robbery. These young men were commonly referred to as "highwaymen".

Over the past 200 years, the gang situation in our country has undergone a dramatic transformation. Even more importantly, the development of youth gang activity during the past 20 years has been profound. Youth gang activity has spread to many regions, communities and socioeconomic groups that in the past would have seemed immune to such influence. Youth gang activity is now having a dramatic effect on our streets, businesses, military, and our schools.

Some of our modern youth gangs have grown into powerful criminal organizations. One nationally recognized authority contrasts the net worth of the leader of the Chicago-based Folk Nation to that of computer tycoon Bill Gates. The major youth gangs of today own businesses to allow for money laundering, have openly endorsed candidates for public office (with some success) and have even applied for and been awarded federal grants.

The increased activity of violent extremist groups and cults is also having a dramatic effect on many segments of our society. While these groups normally do not include in their ranks the large numbers of students who belong to youth gangs, they offer the potential for creating havoc on a larger scale in a single incident. An example is the student involved with this type of group who brings an explosive device to a school. According to a number of experts in the field, these groups are currently the most likely to use explosives to commit acts of violence.

As different examples of this type of threat, two men armed with semi-automatic rifles entered a private religious elementary school in a small Alabama city and took a large portion of the faculty and students hostage. Both men were members of a violent extremist group. Even though the school and local police department were woefully unprepared, the incident was finally resolved without any deaths. Another incident in a Western state involving a husband and wife who took over an elementary school resulted in one death and more than 50 students being injured when an explosive device detonated in the school.

Both youth gang activity and violent extremist groups pose increasing challenges to school safety with student involvement in both types of groups growing. It is interesting to note that while youth gang activity is probably growing at a faster rate in larger communities, violent extremist groups are often more prevalent in rural areas. It is important to remember, however, that communities of all sizes are experiencing significant problems with each type of group. Many school systems have problems with students involved with both.

Another important consideration is the fact that in case after case across the country, when a community that has not previously experienced a problem with either of these types of groups sees early indicators and does not deal aggressively enough with them, a major and well-entrenched problem soon develops. Youth gangs in particular can be likened to contagious disease. Local officials do not want to admit that they have the problem when signs appear, allowing the problem to spread, and it never completely goes away. In community after community across the country, the same pattern has repeated itself. No actions are taken to prevent the problem until obvious signs of gang activity appear, community leaders deny that the problem exists and the groups gain a strong foothold.

Communities without signs of gang activity are strongly advised to develop a "gang inoculation plan" to immunize the community from the development of this problem. By informing key members of the community of indicators of gang activity, periodically assessing the community and providing gang prevention education programs to students, gang activity can be spotted and dealt with aggressively when it first appears. If these steps are taken, it may be possible to prevent the problem from ever gaining a significant foothold.

As the first step, key personnel who are in the best position to spot early signs of gang activity should be given gang awareness training and instructed to report observations to a central contact point with the police department normally the most logical contact point. Key personnel usually include but are not limited to: rank and file police personnel, school police officers, investigators, school teachers, school counselors, school administrators, social workers, youth ministers, scout leaders and key officials from any organization dealing with youth in the community. Gang awareness training should focus primarily on the types of youth gangs that are most prevalent in urban communities in the region but should also include some information on groups that are active in other parts of the country.

The growth of the youth gang problem has created an abundance of experienced individuals who have expertise in dealing with gangs. Typically, many of these individuals develop into effective speakers as they begin to get bombarded with speaking requests once they develop a reputation for competency in dealing with gangs. These individuals soon learn how to provide an effective and interesting presentation on the topics of gangs. In most cases such experts will prove to be regional in their area of expertise and in many cases will have limited knowledge about groups in existence in other areas. It is normally not too difficult to locate a knowledgeable person to come into the community and provide current and useful information concerning gangs.

Once key personnel learn to recognize gang activity, they can begin to look for indicators of gang activity in their daily interaction with young people. A series of meetings should then be held to solicit feedback concerning the presence of gang activity in the community. If any indications are present, a strategy should be developed and implemented promptly to prevent the problem spreading.

Even if little or no signs of gang activity are present, a program of gang prevention education should be put into place in the schools. Normally, such action is taken only after the problem is apparent. Dealing with gang activity is a classic area where an ounce of prevention is worth a pound of cure. Programs like the G.R.E.A.T. Program can help to insulate the community from the threat of gang activity and they can also provide students with valuable tools to reduce their chances of being victimized when they travel in other communities with a serious gang problem.

Unfortunately, many communities have not been fortunate enough to have addressed the problem adequately in its early stages. Thus, gang activity is the biggest current threat to our youth in many communities. And, like the drug problem, gang activity will always effect schools in a community where they are prevalent. Fortunately, the school setting is better suited for gang suppression methods than most other areas of the community. Schools typically can exercise much more control than other settings where young people come together.

It may help us to understand what a gang is and how gangs spread and operate before we address methods of dealing with gang activity. Many gang definitions are currently in use with most encompassing the following points:

- A group of three or more persons
- who work together,
- who commit illegal acts,
- have common identifying characteristics
- and have a name for their group.

Why has gang activity spread around the country so much in recent years? One factor has been the glamorization of the "gangsta" lifestyle in television, movies and particularly in music. Certainly, no young person watches a movie or listens to a gangster rap song and immediately takes to the streets on a crime spree. These mediums, however, expose our youth to the concepts of gang lore in an unrealistic fashion. These entertainment media also reinforce the negative lifestyle of those young people already involved with gang activity.

This impact has been most dramatic with young people who have not traditionally been involved with gang activity. Youth from "traditional" families where both parents are present and from affluent households are involving themselves in gang activity in significant numbers.

Another cause for the dramatic spread of gang activity is the "virus effect" of gang members who move from one community to another. Often, a hard core gang member will move from a heavily gang-infested community to one with little or no gang activity. This can occur for many reasons, one of which being when a habitual criminal offender is "sentenced out of state" by courts which employ this technique. When "sentencing an offender out of state" the judge provides a probated sentence for a given length of time (often 8-10 years). The offender can avoid prison by moving to another state, establishing residence and remaining out of the state for the entire period of time. If the offender returns to the sentencing state during the probation period, the full sentence must be served. Unfortunately, one means for the offender to establish residency in another state is to enroll in school where he may begin spreading the gang virus.

Once an experienced gang member relocates to a new "host" city, he or she typically starts a gang in the new community. Being the "big fish in the small pond", it is normally very easy for even a lower-level gang member to impress the young people in the new community. Once one gang is formed and begins preying on area youth, a second gang typically appears in response and the cycle begins.

While the "virus effect" accounts for a large portion of new gang activity in many locations, the "market-share" concept is just as common. In this instance, gangs emulate legitimate business practice by expanding to new locations. Illustrating this correlation, let's consider a city located on Interstate 75. In this city, crack cocaine is being sold by local drug dealers with several poorly organized youth gangs that the local community regards only as "wannabees" who do not seem to pose a major threat. If none of the major youth gangs is active in this community, leadership of these groups is losing money by not participating directly in this particular market. Just as a national fast food chain will open a local franchise when the market is large enough to be profitable, youth gangs will select an individual community to gain a share of that market.

Although the process is sometimes not quite as formal as that used by the fast food chain, it is just as effective. One of the major groups will send an experienced gang member to the town in question. This individual then "upgrades" one or more of the local gangs and shows them how "real gangstas do it". A full blown organized gang will soon be operating.

The market-share concept has demonstrated very clearly and to some extent predictably its patterns of movement across the country. Chicago-based Folk Nation and People Nation have spread effectively southward and westward while the California-based Crips and Bloods have spread northward and eastward. Unfortunately, hundreds of new communities now have affiliates of the above mentioned "big four" youth gangs.

One contributing factor to this rapid growth is the tendency for government officials to underestimate the power, organization and tenacity of the gangs. Another is to take too lightly local gangs and to dismiss their members as "wannabees." Groups such as Folk Nation and the Crips have grown as rapidly as they have because they are effective at exploiting local groups and very adept at expanding into new markets. Of course, some communities are more susceptible to the rapid spread of gang activity than others. Even with the differences between communities, however, a key factor is normally the manner in which the community deals with the gang threat in the early stages. If aggressive and comprehensive measures are taken quickly, the gang may decide that an easier target may be acquired in another community not so well prepared.

There are many myths concerning gangs and gang members:

- Gang members are ethnic minorities from poor neighborhood. In many areas of the country a high percentage of gang members may fit this general profile, but this is a very dangerous and common stereotype. There are gang members of all races and socio-economic backgrounds. There are gangs made up entirely of white males (or females) all from affluent families. Although these individuals may not be the "typical" gang member it is more and more difficult to define what the "typical" gang member is. The more developed the gang problem is in a community, the broader range of young people will be involved. We must avoid stereotypes that lump young people who are simply different in with gang members.

- Gangs are an urban problem. Just prior to revising this book the author taught at a conference in Montana. One participant reported intensive gang activity in an extremely rural Native American community. Despite incidents of extreme violence, including a juvenile homicide, tribal leaders refused to acknowledge that a gang problem exists. Gangs have become a major problem in suburban and rural communities across the nation.
- Gang members are males. Many youth gangs allow female gang members, and there are a number of youth gangs that consist entirely of females. Such female gang members have been involved with robberies, drug sales, drive-by shootings and murders.
- Gang members are stupid. While many gang members drop out or are expelled from school, good grades are not always a reliable indicator of a child's gang membership. The author recently participated in an arrest of gang-member student on a gun charge. The young man has made straight A's through the ninth grade and has been recognized for academic excellence.
- Gangs are always territorial. This myth has its origins in the traditional Hispanic street gangs. While these groups originally established static and clearly defined territories, many modern gangs have no geographical boundaries.
- Gang graffiti is just a form of harmless street art. For gangs that use graffiti it is an important means to communicate. Graffiti is utilized to mark territory, insult and threaten rival gangs and to brag about recent gang killings to intimidate and advertise criminal activity. Many murders have resulted from gang graffiti. Allowing graffiti to remain sends a message to young people that the traditional authority figures are not in charge; the gangs are. When graffiti is allowed to remain in place at a school, students lose confidence in the administration. When graffiti is permitted to remain, it is clear to the gang member that the gang has more authority than the school.
- Gang members are usually armed with machine guns and automatic weapons. While this situation can be encountered from time to time, it is rare. In most areas gang members use whatever weapons are available.
- Gang activity is a police problem. Because gang activity results from a number of societal problems that are beyond the scope of law enforcement agencies to resolve, the police cannot be expected to solve the problem through enforcement action alone. At the same time any successful gang-reduction strategy will fail without a very strong enforcement component. Gang activity is criminal and solutions must involve aggressive enforcement and prosecution efforts.
- Gangs are just a fad. Those who have made this statement have very often regretted doing so after gang activity has been allowed to become well entrenched.

Gang suppression measures in a school setting involve well designed policies that require aggressive and consistent enforcement of criminal violations. Gang membership in and of itself is not normally a criminal violation, but gang members typically must commit criminal acts to meet the objectives of the gang. Aggressive enforcement will normally affect gang members more quickly because they commit criminal acts on a more frequent basis than other students.

Weapons screening procedures are critical for two reasons. First, gang members rely heavily on the use of weapons to intimidate others and for protection from rival gangs. Secondly, students who are afraid of gang victimization will often bring weapons to school for use in "self defense". Again, many school weapons incidents involve students who have not previously been arrested or involved with significant school disciplinary action.

Another important measure is the restriction of electronic communication devices. Gang members are often heavily reliant on pagers and cellular phones. Drug deals, sales of stolen goods and even hits on rival gang members are often communicated via pagers and cell phones. Often, the phones are utilizing unlawfully "cloned" numbers. Prohibiting pagers and cell phones in school buildings can prove a serious handicap to the gang member. Weapons screening using metal detectors is often a big help in this area since pagers are often detected by screening. One useful strategy is to charge a recovery fee when these devices are returned to parents after a policy violation.

Any measures that can interfere with the gang member's ability to flaunt gang affiliation while at school will be helpful. This includes regulation of gang-related clothing and paraphernalia. Gang members should not be allowed to wear clearly gang-related items such as plastic beads in gang colors or jewelry with gang symbols. Gang members often wear symbols used by established religions such as the six-point Star of David that is worn by Folk gang members in memory of David Barksdale. Unfortunately, this symbol is identical to the six pointed star worn by many devout practitioners of the Jewish faith. Members of all the "big four" youth gangs (Folk, People, Crips and Bloods) have been known to wear the crucifix as a sign of gang affiliation.

Modes of dress that show gang affiliation should also be restricted when possible. In the author's region gang members of the big four often roll up a right or left pant leg to show gang affiliation. School systems that allow this type of behavior on school property are sending a clear message to their youth. Allowing students to wear bandannas on school property can also send a very clear message to gang members and non-gang members alike: the principal is not in charge of the school; gang members are.

A clearly defined dress code that requires that shirts be tucked in at all times and prohibits clothing with large written messages and symbols can help reduce gang problems while still providing significant freedom of dress for students. Gang paraphernalia should also be clearly prohibited along with any gang symbols or other such writing on book bags and clothing. Even though these measures designed to reduce gang identifiers may not be as effective as the more restrictive uniform or "positive dress code" policies, they can bring gang dress down to more manageable levels.

School uniforms with strict color codes, prohibition of all jewelry and requiring that all tattoos be covered are the only means that can effectively neutralize the use of gang identifiers in the school setting. Even then, other restrictions will be required as gang members try to find creative ways to show affiliation-painting fingernails in gang colors, gang-style haircuts, and shaving of eyebrows such as the popular three slashes cut into the right eyebrow to represent the Folk Nation pitchfork. In many school systems strict dress codes will not be practical and a middle ground must be reached. Of course, careful legal research is required before implementing major changes in dress code.

The important key point is to make certain that students realize that the school has taken reasonable measures to minimize the presence of gang identifiers on school property. This shows students that even though gang members can still be spotted, the school still maintains a significant level of control.

Another key area is strict enforcement of policies prohibiting gang-related speech and hand signs. Allowing students to use insulting gang phrases such as "crab," "flea," "donut" and "slob" will generally result in acts of immediate violence. Allowing gang members to flash hand signs will also result in violent incidents. If state law provides an appropriate code section, the use of citations to appear in court can reduce the use of these triggering actions and help to reduce conflict. For those who consider this response extreme, it is important to note that a very high percentage of gang-related shooting incidents are preceded by these actions. It must always be understood that the gang member is under incredible pressure to respond with violence to any supposed insults by a rival gang member.

Indicators of cult and hate group activity must be similarly regulated in the school setting. Such groups are often addressed by policies that deal with gang activity. Prohibiting symbols, jewelry and other identifiers common for these groups is just as important as dealing with gang identifiers.

In some cases, students associated with these groups should be monitored with even greater concern than the known gang member. While the gang member is a high risk individual for gun and knife violations, the violent extremist may be high risk to bring knives, firearms and, of even greater concern, explosive devices into the school setting.

Certain dates can be of higher risk with students who are involved with cult or hate groups, such as Halloween for the student involved with Satanism or the anniversary of the FBI response at the Branch Davidian Compound for the violent extremist militia member. Because these and other dates are often selected for activity by these groups, schools with known extremist activity may need to exercise extra caution on such dates.

Just as gang activity can be glossed over by school or law enforcement officials, so too can activity by these groups. Consistent sanctions for any prohibited activities is recommended. Allowing them to operate in the school setting may create a dangerous environment where violent acts are more likely to occur.

Much of the concern posed by these groups is due to the danger posed by adults who have in several cases committed major acts of violence on school campuses. Probably the most important areas to be addressed are safe school planning and critical incident management. This is particularly true if the school is located in a region where there is a high level of activity involving one or more of these hate groups. Of particular concern is the location of a cult compound in the community. There are a surprising number of these compounds around the country. The author has viewed one such compound (actually a series of pyramid shaped buildings located in a fenced area) in a rural central Georgia community where cult members have been observed to be heavily armed while patrolling the compound. If children who live in this type of environment attend school in the community, the potential for a violent encounter involving an adult cult member needs to be considered. Many of these groups are often prone to home school their children, however, keeping them separate from the mainstream community.

One problem inherent with violent extremist groups and cults is that the individuals who hold their beliefs often do not accept the legitimacy of such government institutions as schools. These individuals have often reacted with acts of extreme physical violence toward what they consider unlawful intrusions by government agencies. Another significant hazard is that these groups may chose an educational facility as a specific target in trying to "hit a nerve" with society in general or to attract media attention. This type of situation has already occurred and is likely to be seen again in the future.

Chapter 11
REFERENCE GUIDE

Following are the names and contact information of individuals and organizations that may provide useful information, training or products to enhance the safety of a school's environment or the efficiency of its school police unit.

- The National School Safety Center is one of the single best sources of information regarding school safety. The center has published numerous books and reports on the topic and produces a newsletter "School Safety Update" six times each year which is available by subscription. The center also has conducted considerable research available on its website. Staff at the center maintain a listing of on all violent school deaths in the United States since 1992. In the author's opinion Executive Editor Dr. Ronald Stephens is the nation's leading expert on school safety and possesses a wealth of information on this field.

The center can be reached at:
Pepperdine University
National School Safety Center
141 Duesnberg Dr.
Suite 11
Westlake Village, CA 91362
(805) 373-9977
Fax: (805) 373-277
www.nssc1.org

- The Office of Juvenile Justice and Delinquency Prevention is a branch of the United States Department of Justice that provides many valuable services. OJJDP produces documents, manuals and literature. The office is an excellent resource for grant funding and for technical assistance. The OJJDP also provides juvenile justice training of exceptional quality. The OJJDP covers the cost of accommodations and the cost of training for a series of courses that are geared toward juvenile services. For the cost of travel and meals, police officers and social service professionals can receive some of the finest training that is currently available anywhere at any price. Course titles include: Child Abuse and Exploitation Investigative Techniques and Responding to Missing and Abducted Children. The OJJDP has also offered a variety of exceptional police management programs that are ideal for the SRO commander or School District Police Chief.

For further information contact:
Office of Juvenile Justice and Delinquency Prevention
810 Seventh Street
NW
Washington, DC 20531
202-307-5911
Fax: 202-307-2093
www.ojjdp.ncjrs.org

- The International Association of Chiefs of Police offers a wide range of training programs that can be helpful to improve the efficiency of the school district police department or SRO program. Training programs are available to members and to non- members (non-members normally pay higher tuition). The IACP also offers a wide range of technical assistance members of the association and the IACP currently administers a Firearms Interdiction and Investigation Grant Program for the Bureau of Justice Assistance.

For further information contact:
The International Association of Chiefs of Police
515 N. Washington St.
Alexandria, VA 22314-2537
(800) THE-IACP
Fax: (703) 836-4543
www.theiacp.org

- The Regional Counterdrug Training Academy at the Naval Air Station in Meridian, Mississippi, is a training facility that offers training for police personnel that is designed to enhance drug enforcement efforts. The facility provides training in a variety of topic areas with several courses that have relevance in the school setting such as Gang School, Interviews, Advanced Interviewing, Statement Analysis and Patrol Officers Response to Street Narcotics. The Center covers all expenses except for travel for all courses that are offered. Priority is given to police officers from Alabama, Louisiana, Mississippi and Georgia but unused slots are often available to officers from other states not in the primary service area.

For further information contact:
The Regional Counterdrug Training Academy
Meridian Naval Air Station
219 Fuller Rd.
Meridian, MS 39309-5020
(877) 575-1435 or (601) 679—2063/3226/3227/3248
Fax: (601) 679-2065
www.rcta.org

- The Federal Law Enforcement Training Center at Glynco, Georgia, offers a wide selection of top quality police training programs in numerous topic areas. Of particular interest to those in the school setting are courses relating to critical incident management. A number of federal agencies provide training to police agencies at the facility.

For further information contact:
The Federal Law Enforcement Training Center
Glynco, GA 31524
(800) 74FLETC
(912) 267-2100
Fax: (912) 267-2894
www.fletc.org

- Every school police unit is urged to obtain a copy of *Safety, Order and Discipline in American Schools*. This manual is revised each year by the Law Advisory Group which is one of the nation's top legal groups in this specialized field.
  To find out about upcoming quality seminars relating to school/police legal matters, or to obtain a copy of the manual contact:

For further information contact:
Law Advisory Group
P.O. Box 19184
Euclid, OH 44119
(888) 529-4770

Fax: (440) 937-2191
www.lawadvisory.org

- Dr. Gary Avery is one of the nation's legal leaders in the field of school police matters. He provides clear, understandable and realistic advise and training that can help school police units solve problems while minimizing civil liability.

Dr. Avery can be reached at:
3907 N. Federal Highway
Suite 118
Pompano Beach, FL 33064
(954) 785-1596

- The National Rifle Association has a variety of educational materials concerning firearms safety including the proven Eddie Eagle Gun Safety Program. This program is focused on reducing gun accidents among children and can be taught to elementary school children in English or Spanish.

For further information, contact:
Community Services Program Division
National Rifle Association
11250 Waples Mill Rd.
Fairfax, VA 22030
(800) 231-0752
(703) 267-1568
Fax: (703) 267-3993
www.mynra.org

- Reid Psychological Systems offers a wide range of written test instruments that are designed and validated to evaluate candidates for employment in terms of honesty, integrity, chemical dependency and other risk factors. The company offers tests that are job specific for a variety of job classifications.

For further information, contact:
Reid Psychological Systems
153 W. Ohio St.

Chicago, IL 60610
(800) 922-7343
Fax: (312) 294-0140

- The Bureau of Alcohol, Tobacco and Firearms administers the Gang Resistance Education and Training Program (G.R.E.A.T.). The BATF can also provide training and offer valuable assistance in firearms incident investigations, tracing of firearms and explosives.

The BATF can be reached at:
Bureau of Alcohol, Tobacco, and Firearms
650 Massachusetts Ave., NW
Washington, DC 20226
(800) 726-7070 or (202) 927-2160
Fax: (202) 927-3180
www.atf.treas.gov/great/atf.htm
or www.atf.treas.gov/schoolsafetyprograms.htm

- The Institute of Police Technology and Management in Jacksonville, Florida, offers a variety of police training courses that may be of interest to the school police officer, such as Hostage Negotiations, Drug Interdiction and Crisis Management.

The institute can be reached at:
Institute of Police Technology and Management
University of North Florida
12000 Alumni Dr. Jacksonville, FL 3224-2678
(904) 620-4786
Fax: (904) 620-2453
www.iptm.org

- The author recommends that school district police chiefs and commanders from large SRO units attend training at one of the three following training programs known as "the big three" in the law enforcement field. Each of these programs lasts approximately three months and is on a level appropriate for those in management or executive positions. All three are considered to be somewhat equivalent in quality even though each program differs. The Federal Bureau of Investigation National Academy Program is probably the best known and has the advantage of having most expenses being covered by the Bureau. With the Bureau spending roughly $20,000 per slot, appointments made by the Director of the FBI are generally difficult to acquire. All three academies also offer shorter training programs in a variety of topic areas.

For further information contact:
FBI National Academy
FBI Academy
Quantico, VA 22135
(703)-640-6131
www.fbi.gov/hq/td/academy/academy.htm

(formerly Northwestern University Traffic Institute)
Northwestern University Center for Public Safety
405 Church St.
P.O. Box 1409
Evanston, IL 60604
(847) 491-5231
(847) 491-5270
http://www.northwestern.edu/nucps/

Southern Police Institute
Department of Justice Administration
College of Arts and Sciences
University of Louisville
Louisville, KY 40292
(800) 334-8635
(502) 852-6561

- LRP Publications offers a periodical relating to school safety, *School Violence Alert*, available by subscription. LRP also offers manuals relating to school safety and periodically offers seminars regarding school safety.

For further information contact:
LRP Publications
747 Dresher Road, Suite 500
P.O. Box 980
Horsham, PA 19044-0980
(800) 341-7874
Fax: (215) 784-9639

- The Georgia Police Academy is located in the Georgia Public Safety Training Center in Forsyth, Georgia. The academy offers a number of courses relating to school safety, including a two-week School Resource Officer training program. Courses are available at no cost to Georgia police personnel and are available for a fee to out-of-state police personnel.

For further information contact:
The Georgia Police Academy
Georgia Public Safety Training Center
1000 Indian Springs Dr.
Forsyth, GA 31029-9599
(912) 993-4000
Fax: (912) 993-4303
www.state.ga.us/gpstc

- The Dogs Against Drugs/Dogs Against Crime (DAD/DAC) National Law Enforcement K9 Association can be a valuable resource to the school district police department or the SRO unit in getting police K9's into the school setting. DAD/DAC is a private not-for-profit organization that aids police agencies which could not otherwise afford police dogs or technical assistance. The organization co-ordinates grants from donations to help police departments purchase dogs. DAD/DAC can also help departments organize fund raisers to pay for dogs. The organization also has performed extensive research in many areas of police K9 information. DAD/DAC can provide information regarding police K9's, equipment and training and can also provide information concerning quality vendors of police dogs and equipment. The organization has created an educational program called The Puppy Patrol that teaches elementary school children about the dangers of drugs. For the institution without funding to buy a gun detection, bomb detection or drug detection dog, DAD/DAC is an invaluable resource.

For further information, contact:
Dogs Against Drugs/Dogs Against Crime
P.O. Box 376
Pendleton, IN 46064
(888) DAD-DAC7
Fax: (765) 642-4899
www.daddac.com

- Multijurisdictional Counterdrug Task Force Training (MCTFT) is operated by the Florida National Guard to provide training focused on narcotics interdiction and related areas of instruction. The academy provides training at numerous sites around the United States. The MCTFT training catalog contains course listings for several topics that would be relevant to the school police officer including criminal street gangs, interviewing and interrogation and patrol drug interdiction. While the author has never attended training through the MCTFT, the academy has a good reputation for providing quality instruction.

For further information, contact:
Multijurisdictional Counterdrug Task Force CDT
3200 34th Street South
St. Petersberg, Fl 3371
(800) 243-5550
Fax: (800) 346-4412

- The School Security Report is a newsletter format periodical that includes a variety of articles concerning school safety issues. The newsletter provides examples of current programs that have proven successful in school systems around the country. It also includes stories pertaining to civil suits against school systems relating to failure-to-protect issues. The newsletter is available by subscription and welcomes information regarding innovative programs relating to school safety. The newsletter also offers a supplemental weekly bulletin listing announcements concerning school safety positions. This service can be of value to those seeking school police management positions and to school systems with a vacant position for a leadership position.

For further information, contact:
Rusting Publications
402 Main St., P.O. Box 190
Port Washington, NY 11050
(516) 883-1440

- The Special Programs Office of York College of Pennsylvania offers a variety of protective services courses that cover topic areas of relevance to the school police officer. Courses offered include Violence Prevention, Interviewing Victims and Witnesses, Crowd Management, Defensive

Tactics, Close Quarters Defense, Officer Survival, Tactical Handcuffing and Public Relations. Courses can also be taught off campus. The author has never attended training through York College, but the program appears to be one of high quality and the topic areas are definitely of relevance to the school police officer.

For further information, contact:
Special Programs Office
York College of Pennsylvania
York, PA 17405-7199
(717) 846-7788
http://www.ycp.edu/

- The Georgia Emergency Management Agency Office of the Governor's School Safety Project employs thirteen full-time school safety experts to assist Georgia School systems and public safety officials through training, technical assistance and crisis response support. GEMA provides numerous free seminars at various locations around the state. The agency also maintains a secure public safety website that is available at no cost to users. The network is now available to police personnel internationally and contains a special school safety section that is open to school officials.

Georgia Emergency Management Agency
Office of the Governor School Safety Project
P.O. Box 18055
Atlanta, GA 30316-0065
http://www.state.ga.us/GEMA
http://safetynet.gema.state.ga.us
(404) 635-7000
1-800-TRY GEMA (in state only)
Fax: (404-635-7005

- The Law Enforcement Television Network offers programs that enable police agencies to train police officers without requiring that the officers travel to a training site. Programs include;

1. Satellites Services, which airs 24 hours per day, five days each week and includes news of interest to law enforcement agencies as well as a wide range of training programs.
2. The Specialized Training, Testing, and Recording System (STARR) Desktop Workstation provides computer-based structured training which allows individual training and testing anytime day or night without requir-

ing supervision of students.

3. The Training on Demand Video Library provides a wide selection of programs that can meet the specific demands of a particular agency.

For further information, contact:
Law Enforcement Television Network
4101 International Parkway
Carrolton, TX 75007
(800) 535-5386
(972) 716-5473
Fax: (972) 716-5252
www.letn.com

- The following resources are compiled from the booklet *Safer Schools - Strategies for Educators and Law Enforcement Seeking to Prevent violence Within Schools*, published by the International Association of Chiefs of Police with funding by the Bureau of Justice Assistance, Office of Justice Programs, U.S. Department of Justice. The author assisted in the development of this publication. It is a an extensive list of agencies and organizations that can provide valuable assistance to the school resource officer.

National Association of School Safety and Law Enforcement Officers (NASSLEO)

- Is a nonprofit organization made comprised of School District Police Chiefs, Directors of Security and Safety, School Police Officers, School Security Officers, School Resource Officers, School Administrators, School Security Consultants and others who work in the field of school safety.

NASSLEO
P.O. Box 25884
Richmond, VA 23260
Telephone/Fax (804) 769-9573
Email: SECNCOMMAN@aol.com
www.nassleo.org

National Association of School Resource Officers (N.A.S.R.O.)

- Is a not a for profit organization, comprised of liaison officers assigned to the school setting. It is the largest training organization for school based police and district personnel in the nation.

N.A.S.R.O.
P.O. Box 40

Boynton Beach, FL 33425-0040
1-800-31NASRO
www.nasro.com

American Association of School Administrators
1801 North Moore Street
Arlington, VA 22209
703-528-0700
Fax: 703-841-1543
Website: aasa.org

Big Brothers Big Sisters of America
230 North 13th Street
Philadelphia, PA 19107
215-567-7000
Fax: 215-567-0394
Website: bbbsa.org

Boys and Girls Clubs of America
1230 West Peachtree St., NW
Atlanta, GA 30309
(404) 487-5700
www.bgca.org

Bureau of Justice Assistance Clearinghouse
P.O. Box 6000
Rockville, MD20849-6000
(800) 688-4252
Fax: (301) 519-5212
www.ncjrs.org

Center for the Study and Prevention of Violence
Institute of Behavioral Science
University of Colorado
Campus Box 439, Building #10
Boulder, CO 80309-0442
(303) 492-8147
Fax: (303) 443-3297
www.colorado.edu/IBS

Community Policing Consortium
1726 M Street, NW, #801
Washington, DC 20036
(800) 833-3085
Fax: (202) 833-9295
Website: communitypolicing.org
Eric Resources Information Clearinghouse
2277 Research Blvd.
Rockville, MD 20850
Mailtstop 4N
(800) LET ERIC

Juvenile Justice Clearinghouse
PO Box 6000
Rockville, MD 20849-6000
800-638-8736
Website:ncjrs.org

National Association of Elementary School Principals
1615 Duke St.
Alexandria, VA 22314-3483
(800) 38NAESP (386-2377)
Fax: (703) 548-6021
www.naesp.org

National Association of Secondary School Principals
1904 Association Drive
Reston, VA 20191
703-860-0200
Fax: 703-476-5432
Website. Nassp.org

National Association of Police Athletic Leagues
618 North US Highway 1 Suite 201
North Palm Beach, FL 33408
(561) 844-1823
Fax: (561) 863-6120
www.nationalpal.org

National Center for Conflict Resolution Education
Illinois Bar Center
424 S. Second St. Springfield IL 67201
(217) 523-7056
Fax: (217) 523- 7066
www.nccre.org

National Clearinghouse on Alcohol and Drug Information
11426-28 Rockville Pike
Rockville, MD20852
(800) 729-6686
Fax: (301) 468-6433
www.health.org

National Clearinghouse on Families and Youth
P.O. Box 13505 Silver Springs, MD 20911-3505
(301) 608-8098
Fax: (301) 608-8721
www.ncfy.com

National Crime Prevention Council
1000 Connecticut Ave., NW
Washington, DC 20036
(202) 466-6272
Fax: (202) 296-1356
www.ncpc.org

National Injury Control and Prevention Center
Centers for Disease Control and Prevention
1600 Clifton Rd., NE
Atlanta, GA 30333
(800) 311-3435
Fax: (404) 639-1623

National Institute for Dispute Resolution
1726 M Street, NW, Suite 500
Washington, DC 20036
202-466-4764
Fax: 202-466-4769

Website: www.ncl.org/anr/partners/nidr.htm

National PTA
330 North Wabash Ave, Suite 2100
Chicago, IL 60611
(800) 307-4PTA
(312) 670-6782
Fax: (312) 670-6783
www.pta.org
National Peer Helpers Association
P.O. Box 32272
Kansas City, MO 64171
(877) 314-7337
Fax: (866) 314-7337
www.peerhelpers.org

National Youth Gang Information Center
Institute for Intergovernmental Research
PO Box 12729
Tallahassee, FL 32317
850-385-0600
Fax: 850-386-5356
Website: iir.com/nygc/

Safe and Drug-Free Schools Program
Office of Elementary and Secondary Education
U.S. Department of Education
400 Maryland Ave, SW
Washington, DC 20202-6123
(202) 260-3954
Fax: (202) 260-7767
www.ed.gov/offices/OESE/SDFS

Streetlaw, Inc.
1600 K St. NW Suite 602
Washington, DC 20006-2902
202-293-0088
Fax: 202-293-0089
Website: streetlaw.org

Teens, Crime, and the Community
c/o Streetlaw, Inc.
1600 K St. NW Suite 602
Washington, DC 2006-2902
(202) 293-0088
Fax: (202) 293-0089
www.nationaltcc.org

U.S. Department of Education
Special Education and Rehabilitative Services
Room 3131 Mary E. Switzer Building
Washington, DC 20202-2524
Website: ed.gov/offices/OSERS/
OSEP/earlywrn.htlm

Youth Crime Watch of America
9300 South Dadeland Boulevard
Suite 100
Miami, FL 33156
305-670-2409
Fax: 305-670-3805
Website: ycwa.org

Youth info (website on adolescence-related issues)
U.S. Department of Health and Human Services
www.youth.os.dhhs.gov

140 Reference Guide

BACKGROUND INVESTIGATION BOOKLET

The following background investigation booklet was adapted by the Bibb County Board of Education Campus Police Department from a document provided to the department by the University of Georgia Police Department. This booklet dramatically improved the background investigation process utilized in Bibb County.

Be sure to review current laws in your state regarding privacy issues before making changes to your background investigation process. Due to the sensitive nature of the school law enforcement role, it is the author's firm belief and experience that a more thorough background investigation is required for officers who will serve in this capacity.

The responses in the booklet should be compared to the results of other investigative steps such as interviews, criminal history, driver history, credit history, interviews with neighbors, polygraph, and integrity test results. Any falsifications by the applicant should cause the hiring agency to seriously consider disqualification of the candidate.

The booklet also serves to indicate how thorough and how efficient the applicant will be in completing required paperwork if hired by the agency. An applicant who is careless in filling out the booklet to obtain a position is likely to be even more careless in doing the work that the position requires if they are selected.

Agency name
Agency address
Agency phone number and web site

Pre-Employment Background Check Instruction Sheet

This application is not an offer or contract for employment. The completion of this booklet or any other screening document does not serve as an agreement to hire you. Any statement to the contrary by any departmental employee is not valid. All information written in this pre-employment background investigation booklet will be used only to determine the suitability and qualifications of the applicant for employment reasons.

1. Print or type in black ink only.

2. Complete all questions in detail where explanations are necessary.

3. Any questions/sections not pertaining to you, list as "N/A".

4. If more writing space is needed throughout the booklet, use the pages provided in the back of the booklet, listing the applicable page and section.

5. You must provide all supportive documents required by the department.

6. Failure to complete the background booklet in its entirety will disqualify you from the hiring process.

Important: Truthful and complete responses to the questions in the booklet are a necessity. Discovery of intentional omissions or false answers will be a basis for your disqualification as an applicant. If an applicant is selected for employment and after employment it is determined that the applicant made intentional omissions or provided false answers, grounds for termination from employment will exist.

The information in the booklet is subject to confirmation by administrative investigation, integrity testing, polygraph, and other forms of testing and investigation.

TELL THE TRUTH, THE WHOLE TRUTH,
AND NOTHING BUT THE TRUTH

Supportive Document List

You will be required to provide the following supportive documents to facilitate the process or your applications. These documents must be turned back in with the completed booklet.

Each document must be an authenticated and legible duplicate of the original. The original must be maintained by the candidate for inspection and verification by the background investigator.

Provide The Following
(if they apply to you)

- Original and three copies of your current driver's license issued by the state where you legally reside.
- Original and three copies of your social security card.
- Original or certified copy and two copies of your birth certificate.
- One copy of all of your marriage licenses and one copy of your divorce papers.
- Copy of any civil or criminal litigation (court cases, law suits, bankruptcies, etc.).
- Original and two copies of your high school diploma/GED diploma.
- Original and two copies of the long form of your DD214 form.
- Two copies of your college transcript(s).
- Two copies of your college diploma(s).
- Two copies of any law enforcement training certificates that you have received.

••WE WILL NOT MAKE COPIES FOR YOU••

Personal History Statement

Position applied for:

____

Name:

____
(Last) (First) (Middle)

List any other name(s) that you have been known by, including previous marriages, name changes, nicknames, etc.

____
____
____
____

Home address:

____
(Number, Street, Apartment Number)

____
(City) (State) (Zip Code)

Home Phone Number: (____) - ____________________

Work Phone Number: (____) - ____________________

Other Phone Number: (____) - ____________________
(Specify) ______________

Date of Birth: ______________________
(Month, Day, Year)

Place of Birth:

____
(City, County, State, Country)

Social Security Number: ______-________-__________

Height: __________ Feet_________ Inches

Weight: ______Pound Hair Color: ____________ Eye Color:_________

Scars/Marks/Tattoos (Indicate type and location on body)

____

____

____

____

Other Identifying Characteristics:

____

(1) Personal History (Continued)

Are you a citizen of the United States? ❑ Yes ❑ No

Are you? ❑ Natural Born ❑ Naturalized

Do you hold dual citizenship? ❑ Yes ❑ No
If so, with what nation?

____

Have you ever renounced your U.S. citizenship? ❑ Yes ❑ No

List all clubs, organizations, memberships and associations which you are or ever have Been a member of or associated with:

____

____

____

What are your hobbies, special skills, and abilities (including speaking foreign languages) which may be used in the position for which you are applying?

____

____

____

____

List any equipment, tools or special devices you can operate which would be useful to the position for which you are applying (including computers, calculators, typewriters, etc.):

____

____

____

Are you related to any person who is the owner of a Bail bonding company, or have you ever worked for any one in the bonding business?
❑ Yes (explain below) ❑ No

____

____

____

Education

Circle the number which represents the last year of formal education you completed:

1 2 3 4 5 6 7 8 9 10 11 12

Are you a high school graduate? ❑ Yes ❑ No

Do you possess a High School Equivalency (GED) Certificate?
❑ Yes ❑No

(2) Personal History (Continued)

If you graduated from high school, list the name of the school, location, (city and state) and the year you graduated. Also provide the name you graduated under.

High school

____
(Name) (City/State)

Address:

____

Year of graduation:_______________

Name used: ______________________________

If you received a GED Certificate, provide the name of the issuing authority, name used and certificate number (if applicable)

____
(Issuing authority) (Number) (Name used)

Indicate below the schools you have attended, location (by city and state) and the years you attended, even if you dropped from the roll without completing that school. Do not include schools that you attended for job proficiency.

Grammar school: Dates attended

Address

Grammar school Dates attended

Address

Junior high/middle school: Dates attended

Address

Junior high/middle school: Dates attended

Address

High school: Dates attended

Address

High school: Dates attended

Address

College/vocational school: Dates attended

Address:

Degree attained: Major field of study:

College/vocational school : Dates attended

Address:

Degree attained: Major field of study:

College/vocational school: Dates attended

Address:

Degree attained: Major field of study:

(3) Residences

List your complete previous addresses (include city, state, and zip code) for the past ten (10) years starting with your present address at the top:

Address

From (Mo/Yr): To (Mo/Yr)

Address

From (Mo/Yr): To (Mo/Yr)

Address

From (Mo/Yr): To (Mo/Yr)

Address

From (Mo/Yr): To (Mo/Yr)

Address

From (Mo/Yr): To (Mo/Yr)

Address

From (Mo/Yr): To (Mo/Yr)

Address

From (Mo/Yr): To (Mo/Yr)

Address

From (Mo/Yr): To (Mo/Yr)

Address

From (Mo/Yr): To (Mo/Yr)

Address

From (Mo/Yr): To (Mo/Yr)

(4) Special Section for Police Officer Applicants

If you are applying for the position of Police Officer, please complete the following section:

Have you ever attended a state certification law enforcement academy?
❑ Yes ❑ No

If yes, certification number: ____________________
Date of graduation: ________________
Academy attended____________________________
Type of certification: ______________

Has your certification ever been suspended or revoked?
❑ Yes ❑ No

If yes, why?

____
____
____

Have you ever been placed on probationary status by the certifying agency?
❑ Yes ❑ No

If yes, why?

____

Have you ever been suspended while a law enforcement officer?
❑ Yes ❑ No

If yes, why?

____
____
____

Have you ever been terminated from a law enforcement agency?
❑ Yes ❑ No

If yes, why?

____
____

Do you object to carrying a weapon (gun) with you, even off duty?
❑ Yes ❑ No

Do you object to working nights, weekends, holidays, and/or on short notice overtime?
❑ Yes ❑ No
If yes, why?

____

____

Have you ever been the focus of an internal/departmental investigation?
❑ Yes ❑ No

If yes, what was the nature of the complaint/investigation?

____

____

____

What were the official results, findings of the investigation?

____

____

____

If the subject of multiple internal investigations, please list all on continuation pages

Have you ever been named in a civil action (suit) for a duty- related incident?
❑ Yes ❑ No

If yes, what were the circumstances?

____

What is the status or what was the outcome of the case?

____

____

____

(5) Law Enforcement Training

Please list below any advanced/specialized law enforcement training that you have completed. You will need to turn in copies of training certificates or transcripts for all listed training when you return this booklet:

Course: length:

Academy/institution:

Course: length:

Academy/institution:

Course: length:

Academy/institution:

Course: length:

Academy/institution:

Course: length:

Academy/institution:

Course: length:

Academy/institution:

Course: length:

Academy/institution:

Course: length:

Academy/institution:

Course: length:

Academy/institution:

Author's note: provide at least two full pages for the candidate to list training attended.

(6) References

Please provide in the spaces below, the names, phone numbers, addresses, and other requested data of six (6) work related references. These references must not be relatives. These people will be asked to appraise your character, abilities, personality, and other job related qualities. Please list persons that can be readily contacted. If a reference works on an evening or a night shift, please indicate this next to their home phone number.

Name: Occupation:

Address:

Home phone: Work phone:

Work relationship: Years known

Name: Occupation:

Address:

Home phone: Work phone:

Work relationship: Years known

Name: Occupation:

Address:

Home phone: Work phone:

Work relationship: Years known

Name: Occupation:

Address:

Home phone: Work phone:

Work relationship: Years known

Name: Occupation:

Address:

Home phone: Work phone:

Work relationship: Years known

Name: Occupation:

Address:

Home phone: Work phone:

Work relationship: Years known

Author's notation: It is suggested that a separate written appraisal form be mailed to each reference to be filled out by the reference and returned. An interview should be conducted to confirm the written appraisal.

(7) Personal references. Please list three (3) personal references, who are not related to you that have known you for at least five years:

Name: Years known:

Address:

Home phone: Work phone:

Occupation/Business/Profession:

How do you know this person?

Name: Years known:

Address:

Home phone: Work phone:

Occupation/Business/Profession:

How do you know this person?

Name: Years known:

Address:

Home phone: Work phone:

Occupation/Business/Profession:

How do you know this person?

Have you applied for a position with our agency before? ❑ Yes ❑ No

Are any members of your family employed with (insert name of school district, municipality, or county government)? ❑ Yes ❑ No

If yes, what department?

Are you now or have you ever been, engaged in any business as an owner or partner, or corporate member? ❑ Yes ❑ No

If yes, give details:

Why did you or why will you leave your last job?

Why do you want to work for our organization?

How many times have you received written reprimands from supervisors/employers for not doing your job correctly or for conduct violations?

Circle the number of times: 0 1 2 3 4 5 More

How many times have you received a warning for being late or absent from work?

Circle the number of times: 0 1 2 3 4 5 More

(9) Work History Continued...

Have you ever been the subject of an internal investigation by an employer (other than any investigations already listed under law enforcement work history)?

❑ Yes ❑ No

If yes, explain:

____

____

____

____

Have you ever been terminated from any job? ❑ Yes ❑ No

If yes, please explain the situation(s):

____

____

____

____

Have you ever been asked to resign in lieu of being terminated?

❑ Yes ❑ No

If yes, please explain the situation(s):

____

____

____

____

Were you ever fired or penalized because of an accident that you caused or were involved in while on the job? ❑ Yes ❑ No

If yes, please explain the situation:

___

___

___

Please list below any awards, commendations or other official recognition that you have received from employers or in connection with your work:

___

___

___

___

Please list all jobs that you have held within the past ten (10) years. if you have been in your current position for ten (10) years of more, please list the 3 jobs that you held prior to your current position. List your present position first. Include any periods of military service and any periods of unemployment. Use the proper time sequence, listing even temporary and part time jobs, no matter how long you worked the job.

Name of employer/business:

___

Street address:

___

Your title/duties:

___

This job was: ❑ Full time ❑ Part time ❑ Temporary

Your supervisor's name: Phone number

___

Reason for leaving:

___

Beginning salary: $________per:________
ending salary: $________per:________

Name of employer/business:

___

Street address:

___

Your title/duties:

___

This job was: ❑ Full time ❑ Part time ❑ Temporary

Your supervisor's name: Phone number

___

Reason for leaving:

___

Beginning salary: $_________per:_________
ending salary: $_________per:_________

Name of employer/business:

___

Street address:

___

Your title/duties:

___

This job was: ❑ Full time ❑ Part time ❑ Temporary

Your supervisor's name: Phone number

___

Reason for leaving:

___

Beginning salary: $_________per:_________
ending salary: $_________per:_________

Name of employer/business:

____

Street address:

____

Your title/duties:

____

This job was: ❑ Full time ❑ Part time ❑ Temporary

Your supervisor's name: Phone number

____

Reason for leaving:

____

Beginning salary: $__________per:__________
ending salary: $__________per:__________

Name of employer/business:

____

Street address:

____

Your title/duties:

____

This job was: ❑ Full time ❑ Part time ❑ Temporary

Your supervisor's name: Phone number

____

Reason for leaving:

____

Beginning salary: $__________per:__________
ending salary: $__________per:__________

Author's note: provide space for at least 12 jobs

(11) Military History

If you were never in the active military, reserves, or national guard, you may write "Not Applicable" at the top of this page and skip the rest of the questions in this section.

Are you now a member of any military organization? ❑ Yes ❑ No

Identify which branch(es) you were enlisted in or a member of:
❑ Army ❑ Air Force ❑ Marines ❑ Navy ❑ Coast Guard

Identify what component(s) of the military you were enlisted in:
❑ Regular ❑ Active Reserve ❑ National Guard

Giver period(s) of military service:

____

From to

____

branch highest rank held type of discharge

____

What was your service number?

List decorations, awards, medals, and citations that you received while in the military:

____

____

Were you ever listed as AWOL (absent without leave)? ❑ Yes ❑ No

Were you ever court martialed, tried on charges, or punished under the Uniform Code of Military Justice including Article 15, Captain's Mast, Office Hours, Page 11, Deck Court, Company Punishment, or any other disciplinary action while a member of the Armed Forces, National Guard, or Reserves?
❑ Yes ❑ No

If yes, explain:

____

____

(12) Military History Continued

Were you ever reduced in rank? ❑ Yes ❑ No

If yes, explain:

Did you ever commit a criminal act while off duty in the military service?
❑ Yes ❑ No

If yes, explain:

Were you ever a member of any foreign armed forces? ❑ Yes ❑ No

If yes, where?

Financial History

A thorough credit history will be obtained on each applicant. The credit standing of each applicant will be considered on an individual basis. Please provide the following information concerning your credit.

What income other than salary do you receive at the present?

How many persons are dependent upon you for financial support (including natural children, stepchildren, and other relatives whether they live in your household or not)?

Do you pay child support? ❑ Yes ❑ No

If yes, what amount and how often? ___________________

Are you currently paying debts ordered by any court (excluding child support)? ❑ Yes ❑ No

If yes, please explain

____

____

____

Do you own or are you buying real estate? ❑ Yes ❑ No

Have you ever filed bankruptcy? ❑ Yes ❑ No

If yes, provide the following:

____

Year filed: state/county filed: chapter: date of discharge:

(13) Debts

What time payments do you owe now? Check those that apply:

❑ Debt consolidation
❑ Doctor bills
❑ Clothing bills
❑ Car payments
❑ Furniture
❑ Hospital/clinic charges
❑ Credit cards
❑ Child support
❑ Student loans
❑ Family/ friends / employer
❑ Finance Companies
❑ Dentist bills
❑ Appliances
❑ Bank loans
❑ Court judgments
❑ Delinquent taxes
❑ Home improvements
❑ Jewelry stores
❑ Alimony
❑ Garnishments

Other debts:

____

____

____

What is your monthly rent/mortgage payment?

____

Are you behind on any payments at this time? ❑ Yes ❑ No

Are any creditors pressing you for payments? ❑ Yes ❑ No

List companies, persons, entities, or firms which you owe money. Include any deferred loans, medical bills, child support, alimony, etc.

Name of creditor: city/state:

Type of business/purpose: outstanding balance:

Name of creditor: city/state:

Type of business/purpose: outstanding balance:

Name of creditor: city/state:

Type of business/purpose: outstanding balance:

Name of creditor: city/state:

Type of business/purpose: outstanding balance:

Name of creditor: city/state:

Type of business/purpose: outstanding balance:

Author's note: include at least 20 spaces for applicants to list creditors

(14) Financial History Continued

Have your wages been garnished within the past ten (10) years?
❑ Yes ❑ No

Do you have any accounts/debts assigned to collections now?
❑ Yes ❑ No

Adding all balances of your debts to all persons/creditors, what is your current total indebtedness? $____________________

Have you ever been a plaintiff or a defendant in a civil lawsuit?
❑ Yes ❑ No

If yes, explain:

____

____

____

Are you engaged in a pending lawsuit at this time? ❑ Yes ❑ No

If yes, explain the situation:

____

____

____

Substance Use History

It is a fact that some individuals experiment with different substances at some time during their life. Please describe your use (experimental or otherwise) of any substance which is considered illegal, either in the form of plants, seeds, gases, pills, propellants, liquids, powders, glue or any inhalants.

Describe the method the substance was administered, i.e. if injected, consumed orally, inhaled, or induced into the body by any means for the purpose of "getting high", relieving pain, easing anxiety, depression, inducing sleep, increasing alertness, increasing body mass or strength, and/or for ANY other reason.

In the first column below, describe the substance you used (in either common street name or medical/chemical name). In the second column, give the approximate age when you first used the substance. In the third column, give the age when you last used the substance. In the last column, estimate (to the best of your knowledge) the total number of times you used the substance. ALL ANSWERS WILL BE VERIFIED BY POLYGRAPH AND BACKGROUND INVESTIGATION.

Substance used	age first used	age last used	Approximate No. times used

Author's note: provide at least fifteen lines for the applicant to list substances used

(15) Misuse of Medication

List any prescription drug you have used which was in the form of pills, liquids, or ANY other form which was not prescribed or given to you by a physician. For example, given to you by other family members or a friend's prescription.

Name of drug	Date/age used	estimated No. of times used
Name of drug	Date/age used	estimated No. of times used
Name of drug	Date/age used	estimated No. of times used
Name of drug	Date/age used	estimated No. of times used

Participation in Drug Distribution

Please explain below if you have ever been involved (even once) in the growing, cultivation, manufacturing, packaging, preparation, distribution, selling, assisting in selling, storing, transportation of, or holding of any substance which is considered illegal.

Type of substance	activity	quantity	date/age
Type of substance	activity	quantity	date/age
Type of substance	activity	quantity	date/age

Have you ever been the holder or operator (even temporarily) of funds, monies, valuables, drugs, paraphernalia, apparatus, land, vehicles, planes, boats, or other transportation vehicles that were used directly or indirectly to buy, grow, store, sell, trade, or transport illegal drugs, substances, stolen property, or contraband? ❑ Yes ❑ No

If yes, explain:

____

____

Alcohol Use History

The questions in this section deal with your use of alcoholic beverages. Answer all questions completely and truthfully. All applicants must answer ALL questions.

I normally drink alcoholic beverages on the average of: _____times per ______

Was there a period of time in your life when you drank more than you do now? ❑ Yes ❑ No

If yes, explain:

____

____

____

(16) Have you ever been counseled by a supervisor/employer because of an alcohol-related problem? ❑ Yes ❑ No

If yes, explain:

____

____

____

Within the past ten (10) years, have you ever called in sick on a job because you were intoxicated or had a hangover? ❑ Yes ❑ No

Within the past ten (10) years, have you consumed alcohol during working hours? ❑ Yes ❑ No

If yes, explain:

____

____

____

Have you ever been arrested because of drinking? ❑ Yes ❑ No

If yes, explain:

____

____

____

Within the past ten (10) years, have you been dismissed from a job because of an alcohol-related problem? ❑ Yes ❑ No

If yes, explain:

____

____

____

Have you ever consumed any amount of alcoholic beverage during lunch at work? ❑ Yes ❑ No

If yes, explain:

____

____

____

Gambling

Indicate below what forms of gambling in which you have been involved within the past ten (10) years:

❑ Bets on races (any type)
❑ Slot machines
❑ Card games
❑ Lotteries
❑ Parlay cards
❑ Dice games
❑ Bets on sports
❑ Numbers
❑ Bingo
❑ Other games

How much money have you spent on lottery tickets within the past six months? $____________

Do you now owe any person or firm money for gambling debts?
❑ Yes ❑ No

Have you ever borrowed money to gamble? ❑ Yes ❑ No

(17) Gambling Continued

Have you ever used someone else's money to gamble? ❑ Yes ❑ No

Would you gamble more if you had more money? ❑ Yes ❑ No

Did you ever work for a gambler or racketeer? ❑ Yes ❑ No

Prior Arrests/Charges

Have you ever been arrested, indicted, detained or otherwise charged with a violation of federal, state, county, parish, municipal, or city law or ordinance, including military, foreign or juvenile law? ❑ Yes ❑ No

If yes, give details as follows:

Date of arrest	charge	arresting agency	disposition
Date of arrest	charge	arresting agency	disposition
Date of arrest	charge	arresting agency	disposition
Date of arrest	charge	arresting agency	disposition

Have you ever been questioned as a suspect in any violation of the law?
❑ Yes ❑ No

If yes, explain:

____

____

Has there ever been a warrant issued for your arrest? ❑ Yes ❑ No

If yes, explain the charge, issuing authority, and disposition:

____

____

Do you have any reason to believe that you may be wanted by any law enforcement agency, foreign or domestic? ❑ Yes ❑ No

If yes, why:

____

____

____

Have you ever been placed on probation or parole? ❑ Yes ❑ No

Are you now or have you ever been a member associated with, a foreign or domestic organization, association, movement, group or combination of persons whose policy or ideals advocate or approve the commission of acts of force or violence to deny other persons their rights under the Constitution of the United States, or laws of any state? ❑ Yes ❑ No

If yes, explain:

____

____

(18) Illegal Activity

Below are examples of situations/crimes people may become involved in. If you have ever been involved as an offender or an accomplice in any violation of local, state, or federal law, whether or not you were arrested, detained, or detected, describe the offense. The polygraph examiner will take time to listen to anything you have to say in this area. Circle any of the following activities that you have been involved with:

Burglary	Armed robbery	Kidnapping
Attempted murder	Manslaughter	Rape
Bad checks	Forgery	Indecent exposure
Threats	Auto theft	Vandalism
Felony theft	Bombing	Extortion
Sex with a minor	Credit card fraud	Purse snatching
Theft of mail	Forged/altered ID	Child molestation
Petty theft	Shoplifting	Perjury
Theft from a vehicle	Assault	False ID
Criminal trespass	Changing price tags	Child pornography
Arson	Sale/alcohol to minors	Commercial gambling Prostitution
illegal drug use	illegal drug sales	Theft of cash
Theft from job	gang activity	illegal wiretapping

In your own words, briefly describe how and what illegal activity you were involved in. Explain the circumstances of each situation:

____
____
____
____
____
____
____
____
____

Have you ever been a member of a youth gang? ❑ Yes ❑ No

If yes, what group, when, and in what city/state:

____
____

Have you ever been a member of a militia group that is not run by a government entity? ❑ Yes ❑ No

If yes, explain:

___

___

Are you applying for this position for dishonest reasons? ❑ Yes ❑ No

Have you ever given a false statement to an official or in an official proceeding? ❑ Yes ❑ No

If yes, when?

___

(19) Driver's History

This section pertains to your personal driving history. Please provide all requested information. If you need more space, use the continuation page at the rear of this booklet. An investigation will be made to check for compliance with local and state traffic laws.

Do you have a current, valid driver's license? ❑ Yes ❑ No

The current status of my driver's license is:
❑ Valid ❑ Suspended ❑ Revoked ❑ Other

___

License Number: State of issue:

___

Expiration date: Class:

If your driver's license is restricted, list restrictions, i.e. corrective lenses, etc.

___

___

___

Do you have any other valid licenses? ❑ Yes ❑ No

If yes, give number and state of issue:

___

___

Has any license you had ever been suspended or revoked? ❑ Yes ❑ No

If yes, explain:

____

____

____

List all driver's licenses you ever had issued by an authority other than (Your State)

State Number Last year held

State Number Last year held

State Number Last year held

Driving Record

List below all traffic citations you have received except for parking violations. Also list the state of license you held at the time you were issued the citation.

City/State: Date: Nature of Violation:

State of license: Penalty/Disposition:

City/State: Date: Nature of Violation:

State of license: Penalty/Disposition:

Author's note: provide at least ten places for applicants to list citations.

(20) Accident Involvement

Have you ever struck a person/vehicle/object without stopping and/or reporting the accident? ❑ Yes ❑ No

If yes, when and where: ______________________________

List below, all vehicle accidents you were ever involved in as a driver, whether reported or not.

Approximate Date	Location	Injuries	Were you at fault?
Approximate Date	Location	Injuries	Were you at fault?
Approximate Date	Location	Injuries	Were you at fault?

Have you ever been involved in an auto accident while driving your employer's vehicle? ❑ Yes ❑ No

If yes, place an asterisk (•) adjacent to the applicable accident listed above.

Have you ever received a "Safe Driver" Award? ❑ Yes ❑ No

If so, where and when?

____

Continuation Section

Page: __________ Section: ______________

Remarks:

____
____
____
____

Page: __________ Section: ______________

Remarks:

____
____
____
____

Author's note: Provide several pages of continuation sections.

CONTRACT FOR LIFE

Student:

I ____________________ have read the section relating to weapons in the student code of conduct. I understand that if I am found to possess any item that has been defined by the code of conduct as a weapon while on school property, I will face the consequences outlined in the code of conduct. These consequences may include arrest, suspension from school, assignment to the alternative school, and expulsion from school depending on the type of weapon if I violate this policy. I understand that the policy on weapons applies to weapons on my person, the mere handling of a weapon by a student, as well as to weapons found in book bags, purse, student lockers, and student automobiles. I agree not to have, possess, conceal, or carry any items defined as a weapon on school (property which includes all school sponsored events). I understand that I will be held accountable for any contraband that is found in my possession, in my book bag, purse, locker, or automobile. I understand that the school district conducts random metal detection, locker checks, and other inspections in an effort to protect me from those who may carry weapons to school.

Signature

Date____________.

Parent:

I ______________________________ have read the section relating to weapons in the student code of conduct. I understand that my child will face arrest, suspension from school, assignment to the alternative school, and expulsion from school depending on the type of weapon if they violate this policy. I understand that the school district conducts random metal detection, locker checks, and other inspections in an effort to protect my child from those who might carry weapons to school.

Signature

Date____________.

School Official:

I ______________________________agree to abide by state and federal laws that mandate that certain types of student weapons violations be reported to the police. I also agree to comply with the school district student code of conduct concerning weapons on campus. I will make reasonable efforts to reduce the danger of weapons on my campus as outlined by the school system.

Signature

Date______________.

LAW ENFORCEMENT OFFICER'S READY REFERENCE GUIDE TO SCHOOL MENTAL HEALTH OCCUPATIONS

Law enforcement officers assigned to the school environment must work closely with mental health professionals to be effective in providing a safe school environment. Law enforcement officers should have a basic understanding of the roles and capabilities of the mental health personnel most often involved with interventions involving students and employees.

Psychologist

- **Education** - Typically a licensed Ph.D. Although some school systems utilize masters-level personnel who are working towards a Ph.D.
- **Function** - School psychologists work in elementary and secondary schools or school district offices to help identify and resolve students' learning and behavior problems. They collaborate with teachers, parents and school personnel to improve classroom management strategies or parenting skills, counter substance abuse, work with students with disabilities or gifted and talented students and improve teaching and learning strategies. They may also evaluate the effectiveness of academic programs, behavior management procedures and other programs and/or services provided in the school setting.
- **How Law Enforcement may utilize** - Psychologists should be utilized as an integral part of threat assessment teams, home searches, consulting and testing challenging students.

Counselor

- **Education** - Typically a masters-level professional. Although some bachelors-level personnel are not uncommon in smaller systems. All states require school counselors to hold a school counseling certification.
- **Function** - Help students evaluate their abilities, interests, talents and personality characteristics to develop realistic academic and career goals. Counselors use interviews, counseling sessions, tests or other methods when evaluating and advising students. They operate career information centers and career education programs. School counselors at all levels seek

to help students understand and deal with their social, behavioral and personal problems. They emphasize preventive and developmental counseling to provide students with the life skills needed to deal with problems before they occur, and to enhance personal, social and academic growth. Counselors can also provide special services, including alcohol and drug prevention programs, and classes designed to teach students how to handle conflicts without resorting to violence. Counselors also try to identify cases involving domestic abuse and other family problems that can affect a student's development. Counselors work with students individually, in small groups or with entire classes.

- **How Law Enforcement may utilize** - Counselors should be a part of threat assessment teams. May also be utilized in the home search process, consulting or used to identify at-risk students. Counselors may also be used to aid in discipline and consequence management for students.

Social Worker

- **Education** - Typically a masters-level professional (MSW). Normally has some type of license.
- **Function** - School social workers diagnose students' problems and arrange needed services, counsel children in trouble and help integrate disabled students into the general school population. School social workers deal with problems such as student pregnancy, misbehavior in class and excessive absences. They also advise teachers on methods of coping with problem students. May also work with families and utilize community resources. May provide home visits and case management, community referral, counseling and resource identification. They also try to identify cases involving domestic abuse and other family problems that can affect a student's development.
- **How Law Enforcement may utilize** - Social workers are useful for community resource identification, home searches, threat assessment and consulting.

Psychiatrist

- **Education** - Always a Physician or M.D. Not commonly employed by a school or school system.
- **Function** - Same functions as a typical physician, including the authority to hospitalize individuals and prescribe medication.
- **How Law Enforcement may utilize** - Consulting or referral. However, costs are associated with services.

Source: Sonayia N. Shepherd, MS

GLOSSARY

Assessment Panel - A structured panel utilized to screen potential candidates for a position using a question-and-answer format and/or observations of candidates performing specific tasks which typically includes interaction between each candidate and other individuals.

Background Check Booklet - A booklet containing numerous questions to be answered by applicants for a law enforcement position. The answers in the booklet are compared to answers provided during other screening steps such as polygraph and integrity testing. The booklet also requires documentation for vital records such as birth certificate and military discharge paperwork.

Background Screening - A process of screening potential candidates to select only those who will perform adequately. Particular attention is focused on screening out candidates who possess the high degree of integrity required for certain sensitive positions such as law enforcement officer.

Battery - A Georgia Code Section misdemeanor-grade assault charge of a high and aggravated nature which involves the intentional use of force to cause substantial physical harm or visible bodily harm to another person. Another similar Georgia Code Section is Simple Battery which involves physical contact of a provoking or insulting nature, or causing minor physical injury.

Conflict Resolution - A skill area that includes the avoiding and successfully resolving situations involving conflict between individuals.

Contraband Screening - Specific techniques designed to deter and detect the presence of contraband on or near a campus.

Contributing to the Delinquency of a minor - A Georgia misdemeanor grade violation that occurs when a person knowingly and willfully encourages, causes, abets, connives, or aids a minor in committing a delinquent act.

Countermeasures - Specific tactics or actions designed to deter specific undesirable behaviors.

Crime Prevention - The practice of preventing crime rather than reacting after it occurs.

Criminal History Check - A review of all prior criminal arrest reports of a candidate for a law enforcement position.

Criminal Trespass - A Georgia Code Section misdemeanor grade violation that can be charged when an individual: refuses to leave the property of another after being told to do so, returns to such property after being told not to do so by an authorized representative, enters property with the intent to commit a crime or intentionally causes less than $500 damage to property belonging to another.

Crisis Intervention -The process of defusing and stabilizing a crisis.

Disguised Weapons - Manufactured or improvised weapons designed in an attempt to deceive others into thinking that the weapon is an innocent item. A few common examples would include: knives that are constructed to look like lipstick containers, ballpoint pens, hairbrushes, etc.

Disrupting a Public School - A Georgia Code Section misdemeanor grade violation of a high and aggravated nature that occurs when an individual disrupts or interferes with the operation of any public school.

Driver History Check - A review of all prior traffic offenses charged to a candidate for a law enforcement position.

Drug Identification - Skill in recognizing controlled substances when observed in the field.

Drug Screen - A laboratory test of a urine sample, used to determine if a job applicant has used a prohibited drug recently.

False Imprisonment Under Color of Law - An unlawful arrest made by an individual who has the power to make arrests under the authority of a government body. A very serious criminal violation in most states. Such arrests may also expose the officer and employing agency to significant civil liability.

Felony - In most states, a crime punishable by more than 12 months of imprisonment. A serious grade of criminal violation. Conviction of a felony can typically result in the loss of certain rights such as the right to vote and the right to own a firearm.

Fitness for Duty - A condition of an officer being physically and mentally prepared to perform his or her duties at a satisfactory level.

Gang Indicators - Observable traits that are frequently utilized by criminal youth gangs to identify themselves to other gang members. These often include normally innocent items that are also worn by the general public.

Impact Device - A piece of police equipment that is designed to incapacitate an assailant through impact. Examples would include standard police batons, side-handled police batons and the more popular collapsible batons.

Inciting to Riot - A Georgia misdemeanor grade violation that occurs when an individual does an act or engages in conduct which urges, counsels, or advises others to riot, at a time and place and under circumstances which produce a clear and present danger of a riot.

Integrity/Honesty Test - A written test instrument that utilizes carefully prepared multiple choice questions to gauge the level of honesty, integrity, and quality service attitude of candidates. Good quality test instruments of this type are scientifically evaluated for accuracy before being sold to agencies.

Judgmental Firearms Training - Training designed to help ensure that a law enforcement officer has a clear understanding of when deadly force can and should be lawfully employed. Such training often includes: classroom instruction, written testing, traditional firearms range training, firearms training involving moving and/or turning targets or targets that pop in and out of view quickly, paintball guns, and video and or computer-projected scenarios with advanced scoring capabilities combined with laser equipped firearms.

Law-Related Education - Programs in which students are trained in the role of the criminal justice system, students rights as victims, the consequences of a conviction in criminal courts and specific criminal laws and traffic codes. These programs are usually taught by teachers, attorneys or law enforcement officers.

Loitering in a School Safety Zone - A Georgia Code Section misdemeanor-grade violation that occurs when a person loiters within 1,000 feet of a school or refuses to leave the school safety zone when lawfully requested to do so.

Metal Detection Screening Station - A stationary post where individuals are screened by metal detection equipment for weapons or other metallic devices .

Misdemeanor - In most states a criminal code violation punishable by less than 12 months of imprisonment. Generally, a less serious grade of criminal violation.

Misdemeanor of a High and Aggravated Nature - A more serious grade of misdemeanor which can result in fines up to $4,000 above the normal $1,000 limit for misdemeanor fines under Georgia Law.

Obstruction of a Law Enforcement Officer - A Georgia Code Section that prohibits anyone from interfering with or obstruction of a law enforcement officer who is in the performance of his or her lawful duties. Normally a misdemeanor-grade violation, but charged as a felony if the offense involves certain types of force used against an officer.

Open Hand Control and Restraint Techniques - Techniques that utilize pressure applied to specific points of the body designed to gain compliance from a resistant individual. To be effective in field situations these techniques require constant practice. Most injuries to law enforcement officers and suspects occur when this means is used to subdue violators.

Pepper Spray - An organic-based irritant spray or foam housed in a canister or other delivery system that is designed to incapacitate an assailant without causing permanent injury. Such sprays utilize Oleoresin Capsicum as the main active ingredient. The foam variety is designed to reduce the problem of innocent bystanders being hit with the spray, the inhaling of residual spray in proximity to the incident or being spread throughout a building by its ventilation system.

Polygraph Testing - A testing process utilizing a machine that measures certain bodily functions which are then interpreted by a trained operator in an effort to determine if the person being tested is providing truthful answers to specific questions.

Possession of Alcohol and Possession of Alcohol for Consumption by a Person Under the Age of 21 - It is a violation of Georgia Law for a person under the age of 21 to possess or possess by having in their body, any amount of alcoholic beverage under normal circumstances.

Psychological Testing - A combination of at least one written test instrument with evaluation and follow up interview by a trained psychologist.

Random Weapons Screening - A system in which individuals are selected for screening through a random numerical sequence. Focused on creating a strong deterrent effect while minimizing inconvenience to those screened and reducing disruption to the academic environment.

Restraining Device - An item of police equipment designed to incapacitate an assailant without using impact as the primary means. Such devices utilize pressure points that would normally be used for open-hand control and restraint techniques.

School District Police Officer - A law enforcement officer directly employed by a school district and assigned to work in the school setting. These officers are often referred to as an School Resource Officer (SRO) or Campus Police Officer.

School Resource Officer - A law enforcement officer who is assigned to work in the school setting. This may be a municipal police officer, sheriff's deputy, state police officer or school district police officer.

Security Retention Holster - Sometimes referred to as a high security holster. A handgun holster that is designed with extra safety security features to make it difficult for anyone other than the wearer to remove the weapon.

Street Officer - A law enforcement officer who performs traditional patrol duties as a primary responsibility.

Unlawful Assembly - A Georgia misdemeanor grade violation that occurs when two or more persons assemble for the purpose of committing an unlawful act and the failure to withdraw from the assembly on being lawfully commanded to do so by a peace officer.

Use of Force - The use of force by a person in a position of authority. Such force can range from verbal commands to open-hand control and restraint, to use of impact weapons or irritant sprays, and culminating with the most serious-the use of deadly force.

Violent Extremist Groups - Groups of individuals who advocate the use of violence against those who do not hold their extreme views or against specific categories of people such as minorities, government officials, or people of a particular religious affiliation.

Visual Weapons Screening - Observing individuals to look for specific, observable, physical actions that are often seen when an individual carries a weapon concealed on their person.

ANNIVERSARY DATES

Important Anniversary Dates for Schools

The following dates are considered to be potentially high-risk dates for major copycat acts of school violence:

October 1 (1997): In a planned shooting, a 16 year old comes to Pearl (Mississippi) High School after he kills his mother at home. He kills two students and wounds seven others with a lever action rifle.

December 1 (1997): A 14 year old student opens fire on a student prayer group at Heath High School in West Paducah, Kentucky; three students are killed and five others are wounded in the attack.

December 5 (1999): In a planned shooting, a 13 year old shoots four students at Fort Gibson Middle School in Fort Gibson, Oklahoma.

February 2 (1996): A 14-year-old youth shoots and kills two classmates and a teacher and wounds others in Moses Lake, Washington in a planned attack.

February 19 (1997): During a planned attack, a 16-year-old student opens fire in the commons area at a high school in Bethel, Alaska and kills the principal and a student. Two other students are wounded.

March 24 (1998): In a sophisticated planned attack in Jonesboro, Arkansas, two boys, ages 11 and 13, shoot and kill a teacher and four girls at Westside Middle School. Eleven others are wounded. One of the attacking students had pulled a fire alarm to draw the victims outside.

April 19 (1993): Federal raid tactical response at the Branch Dividian Compound in Waco, Texas. This date is also the anniversary date for the Oklahoma City Bombing in 1995.

April 20, (1999): Mass weapons assault at Columbine High School. Thirteen victims are killed and numerous others are wounded before the perpetrators commit suicide.

April 24 (1998): A 14-year-old boy kills a teacher and wounds two students at an eighth-grade dance in a planned shooting.

May 1 (1992): A 20-year-old kills four people and wounds 10 others in a siege at his former high school in Olivehurst, California.

May 20 (1999): A 15-year-old student opens fire on classmates, wounding six of them at Heritage High School in Conyers, Georgia. According to prosecutors he admitted to planning the attack "for years."

May 21 (1998): A student murders his parents at his residence and plants explosives devices designed to kill public safety officials (on the 20th). The next day, he goes to Thurston High School in Springfield, Oregon, and shoots 26 people. He tries to kill a law enforcement officer in a detention room several hours after his arrest with a knife that he has strapped to his leg.

These dates are considered to be high-risk dates because of the publicity that they continue to receive in the media and on the Internet. April 19 is considered to be a critical anniversary date for militia groups and those who agree with their beliefs. All dates taken from media website listings (where a copycat violator would likely look them up).

Form for ordering Metal Detector Security Books

Please send the following books:

❑ Effective Security Screening $4.95

❑ School Security Screening $4.95

❑ Effective Ground Searching $4.95

Ram Publishing Company
P. O. Drawer 38649
Dallas, TX 75238
FAX: 972-494-1881
(Credit Card Orders Only)

Handling charges:
Please add $2 for
each book or video
(maximum of $40)

Total for items $__________

8.25% Tax (Texas residents) $__________

Handling Charge $__________

TOTAL $__________

❑ Enclosed check or money order

I prefer to order through

❑ American Express

❑ Discover

❑ MasterCard

❑ Visa

By telephone:
1-800-527-4011

Credit Card Number: ______________________________

Expiration Date: __________ Phone Number (8 a.m.to 4 p.m.) ____________________

Signature (Credit Card orders must be signed.) ____________________

NAME ______________________________

ADDRESS (For Shipping) ______________________________

CITY, STATE, ZIP ______________________________